An Introduction to
Non-Euclidean Geometry

An Introduction to
NON-EUCLIDEAN GEOMETRY

DAVID GANS

New York University

ACADEMIC PRESS   New York and London

A Subsidiary of Harcourt Brace Jovanovich, Publishers

ACADEMIC PRESS, INC.
111 Fifth Avenue, New York, New York 10003

United Kingdom Edition published by
ACADEMIC PRESS, INC. (LONDON) LTD.
24/28 Oval Road, London NW1

Library of Congress Cataloging in Publication Data

Gans, David.
    An introduction to non-Euclidean geometry.

    1.   Geometry, Non-Euclidean.   I.   Title.
QA685.G33   516′.9   72–9326
ISBN 0–12–274850–6

AMS (MOS) 1970 Subject Classifications: 50A10,
50C05, 98A15

To my wife, Charlotte

Contents

ELLIPTIC GEOMETRY

Preface

Although non-Euclidean geometry continues to have a very wide appeal —to the liberal arts student and philosophy major as well as to the specialist in mathematics—there is an extreme shortage of books on the subject suitable as texts for the beginner. It is hoped that the present book may serve such a purpose.

There is so much about non-Euclidean geometry which is unavoidably baffling at the start that every effort should be made, we feel, to present the subject as simply and naturally as possible, without burdening the student with extraneous ideas or unnecessary departures from Euclidean geometry. In our treatment of hyperbolic geometry we have therefore retained Euclid's definition of parallel lines and avoided the introduction of ideal elements. Then, regarding parallel lines as of two types, those with a common perpendicular and those without such a perpendicular, we have organized the whole subject into two main parts. The part dealing with parallels which have a common perpendicular is considerably simpler and shorter than the other and is therefore discussed first, forming most of Chapter III (the chapters preceding this are concerned with preliminary matters, mainly historical). The other part is considered in Chapters IV, V, and VI. This separation of material is achieved by using Saccheri's hypothesis of the acute angle as the hyperbolic parallel postulate rather than the usual statement that through a point not on a given line there passes more than one parallel to the line. Customarily, books make no such separation and begin by discussing parallels which have no common perpendicular, with the result that students have initial difficulty with the subject. To be sure, this is the approach used by the discoverers of non-Euclidean geometry, but it must be remembered that they wrote as men of research and were not concerned with how best to present their ideas to beginners.

In the case of elliptic geometry, partly because of its many subtleties and novelties, partly because suitable discussions of it are even scarcer than of hyperbolic geometry, the need for clear and extended discussions of the subject is particularly great. We have made a special effort to meet this need. In the first place, we treat the two elliptic geometries separately, devoting a chapter to each. Then, in presenting each system care is taken to prepare the ground so that the student may acquire a good understanding of the system before any formal work in it is attempted. To achieve this understanding an informal description of each system is given, preceded in the case of double elliptic geometry by a study of the geometry on a sphere, and in the case of single elliptic geometry by a study of the geometry on a modified hemisphere. The formal work in each system consists of an axiomatic development. These developments, which represent some of the most original aspects of the book, are included only for their educational value and make no claim to being rigorous, definitive presentations.

The arrangement of the book makes it suitable for courses of different lengths and contents, and for students of varying backgrounds. Each of the first six chapters depends on those that precede it, but the seventh and eighth depend only on the first three. Hence, a brief course covering both hyperbolic geometry and elliptic geometry can be built on Chapters I, II, III, and VII. With more time available most teachers will probably wish to include Chapter IV. An even briefer course, dealing only with hyperbolic geometry but offering much that is of value educationally, can be based on Chapters I, II, and III. In fact, the material in these chapters, supplemented by a small part of Chapter VII, has a breadth which could also make it useful in geometry courses not devoted exclusively to non-Euclidean geometry, or even in other liberal arts courses. For the teacher who wishes to do Chapter V, on horo-cycles, we mention that the numerical ideas in the chapter, although of con-siderable intrinsic interest, only reach their culmination in Chapter VI, where they are used in deriving the relations among the sides and angles of a triangle. As for Chapters VII and VIII, on the elliptic geometries, it should be noted that much can be learned about these systems even if the sections on their axiomatic developments are omitted.

The required background for the brief courses we have mentioned is minimal: in most cases a year of college level mathematics, together with an interest in geometry, should suffice. In a course covering all or most of the book the student should have the mathematical maturity usually expected of those who have had a year of calculus. However, no calculus is used in the book.

The writer welcomes this opportunity to express his appreciation to Professor Leroy M. Kelly of Michigan State University for his valuable criticism of the manuscript; to Helen Samoraj for her expert typing of the manuscript; to his wife, Charlotte, for her indispensable editorial assistance; and to the staff of Academic Press for their friendly cooperation.

D. G.

HISTORICAL INTRODUCTION

I

Euclid's Fifth Postulate

1. INTRODUCTION

Until the nineteenth century Euclidean geometry was the only known system of geometry concerned with measurement and the familiar concepts of congruence, parallelism, and perpendicularity. Then, early in that century, a new system dealing with the same kinds of things was discovered. Differing from Euclidean geometry in its basic assumption regarding parallel lines, the new system, which came to be known as *non-Euclidean geometry*, therefore contained many theorems that disagreed with Euclidean theorems. This represented nothing less than a revolution in geometry. In the course of time the effects of the discovery on other branches of mathematics, on physical science, and on philosophy proved to be no less profound.

As is often the case with new ideas, the discovery of non-Euclidean geometry was a by-product of efforts to achieve something seemingly quite different. Starting almost from the time of Euclid (about 300 B.C.), criticism had been directed at one of his postulates which, because of its length and complexity, seemed more like a theorem than a postulate. Attempts were therefore made to prove the postulate. These efforts continued for more than 2000 years and were unsuccessful, but they culminated in the discovery of non-Euclidean geometry. We consider some of these efforts in the next chapter. In the present chapter we mainly examine the postulate and its place in Euclid's work.

2. EUCLID'S STATED ASSUMPTIONS

Euclid states ten assumptions and many definitions as his basis for proving all theorems. These assumptions consist of five postulates and five axioms (or common notions) as follows*:

Euclid's Postulates

Let the following be postulated:

1. To draw a straight line from any point to any point.
2. To produce a finite straight line continuously in a straight line.
3. To describe a circle with any center and distance.
4. That all right angles are equal to one another.
5. That, if a straight line falling on two straight lines (in the same plane) makes the interior angles on the same side less than two right angles, the two straight lines, if produced indefinitely, meet on that side on which are the angles less than the two right angles.

Euclid's Axioms (or Common Notions)

1. Things which are equal to the same thing are also equal to one another.
2. If equals be added to equals, the wholes are equal.
3. If equals be subtracted from equals, the remainders are equal.
4. Things which coincide with one another are equal to one another.
5. The whole is greater than the part.

It is seen that the postulates are assumptions of a geometrical nature, whereas the axioms are more general assumptions, applicable to all of mathematics. Today we make no such distinction and use the words "postulate" and "axiom" interchangeably.

Postulate 1 is usually interpreted to mean that there is one and only one straight line through any two points. The term "finite straight line" in Postulate 2 means straight line segment, and therefore this postulate permits the extension of such a segment beyond each endpoint, without, however, specifying the amount of such extension. In Postulate 3 "distance" means the length of the radius. Postulate 5 is the postulate that was subjected to so much criticism. It is illustrated in Fig. I, 1.† Line k intersects lines g, h,

---

* Taken, with permission, from T. L. Heath, *The Thirteen Books of Euclid's Elements* (Cambridge University Press, New York and London, 1926). We shall refer to this work hereafter as the *Elements*.

† In this book we are concerned only with plane geometry. The word "line" used alone will always mean straight line.

forming interior angles *a*, *b* on the same side of *k*. If $a + b$ is less than two right angles, then, according to Postulate 5, *g* and *h* will meet on that side of *k*.

As we shall see, Postulate 5 plays a decisive role in Euclid's theory of parallels. For this reason it is often called his *parallel postulate*.

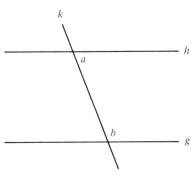

Fig. I, 1

3. THE EXTENT OF A STRAIGHT LINE

In addition to his ten stated assumptions Euclid used many unstated assumptions. By this we mean that in the course of his proofs he took certain geometric properties for granted, never proving them nor stating them explicitly. Just one of these unstated assumptions is relevant to our present discussion. It concerns the extent of a straight line. Euclid regarded a straight line as being infinite in extent. That he did so can be seen, for example, in Postulate 3. It can also be seen in his Proposition 12,* where the term "infinite straight line" is used. Further, in the proof of Proposition 16 he extends a line segment by its own length, thus obtaining a new segment twice as long as the original one. This assumes that any line segment can be doubled. By repeating this doubling one could, of course, obtain a segment whose length exceeds all bounds. In effect, then, Euclid is assuming that a line is infinite. Because of its importance to our subject we shall state Proposition 16 and give what is essentially Euclid's proof.

In all our work $\overline{AB}$ will denote a line segment with endpoints *A*, *B*, and *AB* will denote the length of this segment.

---

* The Appendix of this book lists the definitions, stated assumptions, and propositions in Book I of Euclid's *Elements*. Our subsequent references to the *Elements* are understood to apply only to Book I.

Proposition 16. *In any triangle, if one of the sides is extended, the exterior angle is greater than either of the opposite interior angles.*

Proof. Let ABC be a triangle (Fig. I, 2) and let side $\overline{BC}$ be extended to D. We shall prove that

$$\measuredangle ACD > \measuredangle BAC.$$

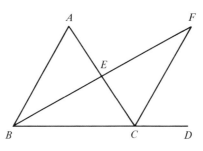

Fig. I, 2

Let $\overline{AC}$ be bisected at E (Prop. 10). Extend $\overline{BE}$ to F so that $BE = EF$. The vertical angles AEB and CEF are equal (Prop. 15). Since triangles AEB and CEF are then congruent by side–angle–side (Prop. 4), we have

$$\measuredangle BAC = \measuredangle ACF.$$

But, the whole being greater than any of its parts (Ax. 5),

$$\measuredangle ACD > \measuredangle ACF.$$

Hence

$$\measuredangle ACD > \measuredangle BAC.$$

Similarly, one can show that $\measuredangle ACD > \measuredangle ABC$.

Euclid uses Proposition 16 directly or indirectly in proving many familiar facts about triangles, for example, that *the greater side lies opposite the greater angle* and that *any two sides taken together exceed the third side*. Proposition 16 is also a cornerstone in his theory of parallels, as we shall presently see.

EXERCISES

1. Finish the proof of Proposition 16 by showing that $\measuredangle ACD > \measuredangle ABC$.

2. Give Euclid's proofs of the following, which depend on Proposition 16:

(a) Proposition 17; (b) Proposition 18;

(c) Proposition 19; (d) Proposition 20.

3. Nothing is lost by ending Postulate 5 with the word " meet " inasmuch as what follows can be proved. Give the proof.

4. EUCLID'S THEORY OF PARALLELS

Euclid calls two straight lines *parallel* if they are in the same plane and do not meet (Def. 23). That parallel lines do exist then follows from Proposition 16. To see this, let g be any line (Fig. I, 3). At each of two* points

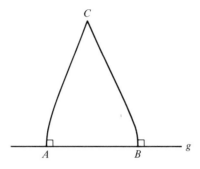

Fig. I, 3

A, B on g let the perpendicular to g be drawn (Prop. 11). If these perpendiculars met in a point C, we would have a triangle ABC such that the exterior angle at B and the opposite interior angle at A, being right angles, are equal. This contradicts Proposition 16. Hence the perpendiculars cannot meet. They are therefore parallel. Instead of using the perpendiculars at A and B, we could have used any lines through A and B such that the so-called corresponding angles a and b (Fig. I, 4) are equal, which is possible according to Proposition 23.

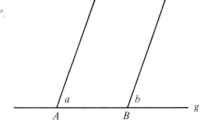

Fig. I, 4

* "Two points," "three points," etc., will always mean two distinct points, three distinct points, etc.

Since Euclid assumed the infinite extent of straight lines in proving Proposition 16, we see that the existence of parallel lines is a consequence of that unstated assumption. Also we see that the statement "parallel lines exist" could have served as Euclid's Proposition 17. However, he delays using the word "parallel" until he reaches Propositions 27 and 28, which state that two lines g, h (Fig. I, 5) are parallel if, when cut by a transversal t,

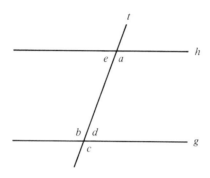

Fig. I, 5

the alternate interior angles a, b are equal, or the corresponding angles a, c are equal, or $a + d$ equals two right angles. The proofs of these propositions are left to the reader.

An immediate consequence of Propositions 27 and 28 is that a parallel to any line g (Fig. I, 6) can be constructed through any point P not on g. We need only take a point Q on g, draw line PQ, construct line h through P so that $a = c$, and then infer that g and h are parallel by Proposition 28. In particular, Q could be the projection of P on g, that is, the foot of the perpendicular from P to g (Prop. 12), in which case g and h would be perpendicular to line PQ.

In proving Proposition 29, which is the converse of Propositions 27 and 28 combined, Euclid used Postulate 5 for the first time. It is therefore clear that he did not need this postulate in order to know that parallel lines exist or, in particular, that a parallel to any line g can be constructed through any point P not on the line. The effect of Postulate 5, as we now show by proving Propositions 29 and 30, is to rule out the possibility that there is more than one parallel to g through P.

Proposition 29. *A straight line falling on parallel straight lines makes the alternate interior angles equal to one another, the exterior angle equal to the opposite interior angle, and the interior angles on the same side equal to two right angles.*

*Proof.** Assume that lines *g* and *h* are parallel, and that they are cut by a transversal *t* (Fig. I, 5). If *a* + *d* is less than two right angles, then *g* and *h* meet to the right of *t* (Post. 5). This contradicts our hypothesis. If *a* + *d* is greater than two right angles, then *b* + *e* is less than two right angles (Prop. 13), in which case *g* and *h* meet to the left of *t*. This contradicts our hypothesis. Hence, *a* + *d* is equal to two right angles. It follows that *a* = *b* and *a* = *c* (Prop. 13).

Proposition 30. *Straight lines parallel to the same straight line are also parallel to one another.*

Proof.† Assume that the distinct lines *h* and *j* are both parallel to line *g*. If *h* is not parallel to *j*, then *h* and *j* meet in a point *P* (Fig. I, 6). Then *a* ≠ *b*

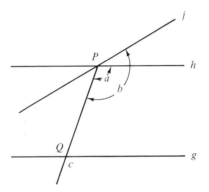

Fig. I, 6

inasmuch as *h* and *j* are distinct. Since *g*, *h* are parallel we have *a* = *c* (Prop. 29), and since *g*, *j* are parallel we have *b* = *c*. Hence *a* = *b*. From this contradiction we infer that *h* is parallel to *j*.

It follows from Proposition 30 that through a point *P* not on a given line *g* there cannot be more than one parallel to *g*; for if there were two, they would have to be parallel to each other by Proposition 30, and this would contradict that they meet in *P*. The proof of Proposition 30 shows that the important fact, *there is not more than one parallel to a line through a point outside the line*, is a consequence of Proposition 29 and hence of Postulate 5. In fact, all properties of parallel lines, except the single property used in defining them, are consequences of Postulate 5. For example, it easily follows from

* Our proof is a concise version of Euclid's.
† This proof is different from Euclid's.

Propositions 29 and 30, and hence from Postulate 5, that a line perpendicular to one of two parallels is also perpendicular to the other. From this, with the aid of Proposition 26, one can then show that parallel lines are equidistant from one another.

EXERCISES

Prove the following facts without going beyond Proposition 30 and note which depend on Postulate 5.

1. Through a point not on a given line there always passes a line parallel to the given line.

2. A line that meets one of two parallels also meets the other.

3. Two lines are parallel if they have a common perpendicular.

4. A line perpendicular to one of two parallels is also perpendicular to the other.

5. Parallel lines are equidistant from one another.*

5. FURTHER CONSEQUENCES OF POSTULATE 5

Many notable facts of Euclidean geometry besides the properties of parallel lines are consequences of Postulate 5. Among them are the Pythagorean Theorem, the formulas for the circumference and area of a circle, the fact that through any three noncollinear† points there passes a circle, the existence of similar figures which are not congruent, the formula for the sum of the angles of a triangle, and the numerous consequences of that formula. Since the connection of the postulate with these various situations is not obvious, let us consider a few of them.

Taking the case of the triangle first, suppose that we have a triangle ABC with angles a, b, c (Fig. I, 7). If we draw through A the parallel to line BC,

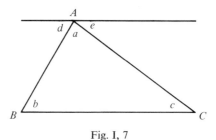

Fig. I, 7

* For our later use we mention that the proof of this depends on Postulate 5.
† "Noncollinear" means "not on a line."

angles *d* and *e* are formed such that $b = d$ and $c = e$ (Prop. 29). Also, $a + d + e$ equals two right angles (Prop. 13). Substitution then gives $a + b + c$ equal to two right angles. This proof of the familiar formula is seen to depend on Proposition 29, and hence on Postulate 5. Any other proof would likewise be a consequence of Postulate 5.

Now let us consider the case of similar figures and prove that there exist similar triangles which are not congruent. By definition, two triangles are *similar* if their angles are equal, respectively, and their pairs of corresponding sides are proportional, corresponding sides being sides that are opposite equal angles. Also, we recall the familiar theorem which states that two triangles will be similar if their angles are equal, respectively. Now take any triangle *ABC* (Fig. I, 8). Let *D* and *E* bisect $\overline{AB}$ and $\overline{AC}$ (Prop. 10). By

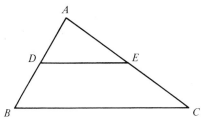

Fig. I, 8

another familiar theorem lines *DE* and *BC* are then parallel. Hence $\angle ADE = \angle ABC$ and $\angle AED = \angle ACB$ (Prop. 29). We conclude that triangles *ABC* and *ADE* are similar. Since $AD \neq AB$, they are not congruent. This proof is seen to depend indirectly on Postulate 5.

To give our final example of properties implied by Postulate 5 let us consider any three noncollinear points *A*, *B*, *C* (Fig. I, 9). Let *g* and *h* be the perpendicular bisectors of $\overline{AB}$ and $\overline{AC}$. To prove that *g* and *h* meet, let us suppose that they are parallel. Line *AB*, being perpendicular to *g*, must then also be perpendicular to *h* (Props. 29, 30). Since lines *AB* and *AC* are

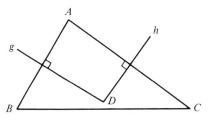

Fig. I, 9

distinct and perpendicular to h, they are parallel (Prop. 28). But this is clearly impossible. Hence g and h cannot be parallel. Their common point D, being equidistant from A, B, C, is therefore the center of a circle through A, B, C. This proof that there exists a circle through any three noncollinear points clearly makes indirect use of Postulate 5.

Any alternatives to the two proofs we have just given would likewise be found to depend on Postulate 5.

EXERCISES

In Exercises 1 to 5 prove the stated facts of Euclidean geometry and show how each depends on Postulate 5.

1. Rectangles exist.

2. Squares exist.

3. An angle inscribed in a semicircle is always a right angle.

4. Opposite sides of a parallelogram are equal.

5. Two right triangles are similar if an acute angle in one equals an acute angle in the other.

In Exercises 6 to 8 show that the stated facts of Euclidean geometry can be proved without using Postulate 5 directly or indirectly.

6. Circles exist.

7. Parabolas exist.

8. Not more than one angle of a triangle can be obtuse.

In Exercises 9 to 11 determine whether the stated facts of Euclidean geometry can be proved without using Postulate 5 directly or indirectly.

9. Equilateral triangles exist.

10. Ellipses exist.

11. The diagonals of a parallelogram bisect each other.

6. SUBSTITUTES FOR POSTULATE 5

The reader has perhaps wondered why he cannot recall Postulate 5 from his previous studies, considering the far-reaching importance of the postulate in Euclidean geometry. Most current textbooks on the subject are adaptations of Euclid's *Elements* to the needs of present-day students and involve simplification, abridgment, and rearrangement. In particular, these texts rarely use Postulate 5, substituting for it some briefer but logically equivalent assumption. The most common assumption of this sort is that *through a point not on a given line there passes not more than one parallel to the line*. This is known as Playfair's Axiom, after the geometer Playfair, who used it in his 1795 edition of the *Elements*. However, he was not its discoverer.

Assumptions equivalent to Postulate 5 play an important part in attempts to prove the postulate and hence deserve our attention. A knowledge of such assumptions, moreover, is useful in appreciating many facts in non-Euclidean geometry. Let us introduce the term *basis E* to mean the set of all definitions and assumptions (stated or unstated) used by Euclid with the exception of Postulate 5. Then, to show that a given statement is equivalent to, or a substitute for, Postulate 5 within the context of Euclidean geometry, we must do two things: (1) prove the statement, using only Postulate 5 and the basis *E* (or their consequences), and (2) prove Postulate 5, using only the statement and the basis *E* (or their consequences). The following is a partial list of such statements.

Substitutes for Postulate 5

(a) Through a point not on a given line there passes not more than one parallel to the line.

(b) Two lines that are parallel to the same line are parallel to each other.

(c) A line that meets one of two parallels also meets the other.

(d) If two parallels are cut by a transversal, the alternate interior angles are equal.

(e) There exists a triangle whose angle-sum is two right angles.

(f) Parallel lines are equidistant from one another.*

(g) There exist two parallel lines whose distance apart never exceeds some finite value.

(h) Similar triangles exist which are not congruent.

(i) Through any three noncollinear points there passes a circle.

(j) Through any point within any angle a line can be drawn which meets both sides of the angle.

(k) There exists a quadrilateral whose angle-sum is four right angles.

(l) Any two parallel lines have a common perpendicular.

We shall prove two of these substitutes now, leave others as exercises, and verify the remaining ones in our study of non-Euclidean geometry.

---

* All perpendicular distances from either line to the other are equal.

Proof of Substitute (a). Since we already know from Section 4 that statement (a) can be proved when Postulate 5 and the basis E are used, it only remains to prove Postulate 5 when statement (a) and the basis E are used. Hence, let g, h (Fig. I, 10) be two lines such that, when they are cut by a transversal PQ,

$$a + b < \text{two right angles.} \tag{1}$$

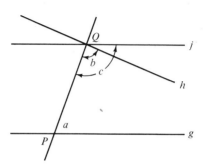

Fig. I, 10

We must show that g, h meet to the right of line PQ. Through Q take line j so that $a + c$ equals two right angles. Then h, j are distinct since $b \neq c$. Also, j is parallel to g (Prop. 28). Because of statement (a), which we are assuming, h cannot also be parallel to g. Hence g and h must meet in some point R. If R were to the left of line PQ, the sum of the angles at P and Q in triangle PQR would exceed two right angles in view of (1). This contradicts Proposition 17. Hence R is to the right of line PQ.

Proof of Substitute (f). We have already seen that statement (f) can be proved when Postulate 5 and the basis E are assumed (§4, Ex. 5). Now we must prove Postulate 5 when (f) and basis E are assumed. Therefore let g, h (Fig. I, 11) be parallel lines and suppose they are equidistant from one

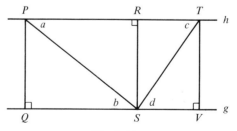

Fig. I, 11

another. Take three points P, R, T on h and let Q, S, V be their projections on g. In the diagram we have taken R between P and T, in which case S is between Q and V. If F, the projection of S on h, were different from R, we would have a triangle FRS with $FS = RS$, and hence with right angles at F and R. This is impossible (Prop. 17). It follows that $\overline{SR}$ is perpendicular to h.

Right triangles PRS and SQP are then congruent, having the hypotenuse and an arm equal, respectively. (The proof of this congruence theorem can be given without using Postulate 5 or any substitute. See Ex. 5.) Hence $a = b$. Likewise, $c = d$ from the congruence of triangles TRS and SVT. From

$$b + \angle PST + d = \text{two right angles}$$

we then obtain

$$a + \angle PST + c = \text{two right angles.}$$

Thus, the angle-sum of triangle PST is two right angles. According to Substitute (e), this implies Postulate 5. There is no circularity in this reasoning, for Substitute (e) can be proved without making any use of Substitute (f).*

EXERCISES

1. Prove Substitute (b).

2. Prove Substitute (c).

3. Prove Substitute (d).

4. Show that statement (j) can be proved if one uses Postulate 5 and the basis E.

5. Using only the basis E prove that two right triangles are congruent if the hypotenuse and an arm of one are equal to the hypotenuse and an arm of the other. Do not use the method of superposition, that is, of placing one triangle on the other.†

6. Assuming Substitute (e) has been proved, show that the statement, *an angle inscribed in a semicircle is a right angle*, is equivalent to Postulate 5.

7. Prove Proposition 29 by using Playfair's Axiom instead of Postulate 5. (In §4 we saw that Euclid used Postulate 5 to prove this proposition.)

* See H. Wolfe, *Introduction to Non-Euclidean Geometry*, pp. 21–23 (Dryden Press, New York, 1945).
† This method is intuitive and we shall avoid it.

Attempts to Prove
the Fifth Postulate

1. INTRODUCTION

The geometers who criticized Postulate 5 over the centuries did not question that its content was a mathematical fact, but only that it was not brief, simple, and self-evident, as postulates were supposed to be. They were familiar with the first three equivalents of Postulate 5 listed in Chapter I, Section 6, which are briefer and apparently simpler than the postulate, and therefore could have rectified matters, it would seem, merely by substituting one of them for the postulate. Some, in fact, did this, but many others sought to prove Postulate 5 on the basis of Euclid's other assumptions, believing that these other assumptions were adequate for the complete development of Euclidean geometry. It was their efforts that eventually led to the discovery of non-Euclidean geometry. Since an understanding of these efforts can cast light on non-Euclidean geometry and also provide new insight into Euclidean geometry, we shall consider some of them in the present chapter.

2. EUCLID'S CHOICES

The task which Euclid set for himself in writing his famous *Elements* was to take the mathematical knowledge that the Greeks had amassed since the time of Thales (640–546 B.C.), bring it up to date with his own contributions, and organize the whole into a single logical system. This involved deciding which properties of geometrical figures were to serve as their definitions, choosing the statements which were to be the basic assumptions of the system and those which were to be the theorems, arranging the theorems in correct logical order, and supplying all the proofs, original ones when necessary. This was an enormous project and, considering the duration and influence of the *Elements*, an eminently successful one.

Euclid's task as a whole is well illustrated by the problem he faced with regard to parallelism. All the facts about parallel lines were at hand, that is, he knew them all: Parallel lines are coplanar; they do not meet; they are everywhere equidistant; a line meeting one of two parallels also meets the other; through a point not on a line there passes a unique parallel to the line; if two parallels are cut by a transversal, the alternate interior angles are equal; and so on. Let us suppose that Euclid had already chosen all of his assumptions except Postulate 5 and was ready to develop his theory of parallels. His first problem was to decide which property of parallel lines to use in defining them. Several choices were open to him. Choosing the familiar definition, the one we have been using, he knew immediately that parallel lines exist, for this is a consequence of Proposition 16, as we have seen. Had he defined parallel lines to be lines that are coplanar and equidistant, which many a student would like to do, he would not have known on the basis of Proposition 16 that parallels with these properties existed, nor would he have known on the basis of any of the first 28 propositions that they existed. This would have been a disadvantage of the definition. Had he defined parallel lines to be lines such that, when cut by a transversal, the alternate interior angles are equal, he could have concluded that parallel lines exist since they would have been easy to construct, and he could have proved, by using Proposition 16, that they do not meet. This definition therefore could have been used instead of the familiar one. But it is clearly not as brief nor as simple.

Euclid's second problem with regard to parallels was to determine whether all the remaining properties of parallels, those not mentioned in the definition, could be proved without introducing new assumptions. We know his decision: One new assumption is needed. In view of the great efforts made by those who tried, unsuccessfully, to prove Postulate 5 on the basis *E*, it is reasonable to conjecture that Euclid did not reach this decision easily.

Finally, Euclid had to decide what the new assumption was to be. Since there can be little doubt that he knew several alternatives to his Postulate 5, we must conclude that he had reasons for preferring it. Today we commonly regard Playfair's statement, *through a point not on a given line there passes not more than one parallel to the line*, as preferable to Euclid's Postulate 5. But it is possible that Euclid looked with disfavor on the idea of using a negative statement like this in the foundation of his mathematical system. It must be remembered that, for Euclid and his contemporaries, axioms and postulates were supposed to be *affirmations* of truth.

3.   POSIDONIUS AND HIS FOLLOWERS

One of the earliest known critics of Euclid was Posidonius (first century
B.C.), who found fault with Euclid's definition of parallels, as well as with
his Postulate 5. Believing that the postulate could be proved with a better
definition of parallel lines, he proposed defining them as coplanar lines that
are everywhere equidistant from one another. As was noted earlier, with
this definition one cannot conclude from Proposition 16 that such parallel
lines exist. In fact, if this definition is used, it is impossible to prove their
existence on the basis of Euclid's assumptions, excluding Postulate 5. We are
in no position now to justify this statement,* but we can try to clarify it by
two examples. Suppose that, at distinct points A, B of a line g (Fig. II, 1),

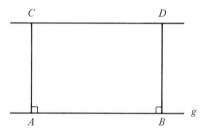

Fig. II, 1

we constructed equal segments $\overline{AC}$, $\overline{BD}$ perpendicular to, and on the same
side of, g, and then tried to prove that lines g and CD are parallel in the
sense of Posidonius's definition. Or, suppose that we considered the locus h
of points on the same side of g which are at some specified distance d from g
(Fig. II, 2) and tried to show that h is a straight line, and hence parallel to g
in Posidonius's sense. Neither attempt could succeed if we used only the
basis specified above.

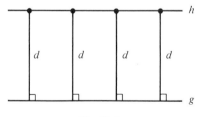

Fig. II, 2

* Our work in Chapters III and IV will help to do so.

It is not known how far Posidonius proceeded with his plan to prove the postulate; we have mentioned him mainly because he was one of the first to criticize Euclid's handling of parallels. But many who came after him attempted to follow the same plan, G. Vitale (1633–1711) being one of the last, and the details of some of their efforts are known. Let it suffice to say that they either gave erroneous proofs that Posidonius's kind of parallels exist or assumed their existence and went on to prove the postulate. As we know from Chapter I, Section 6, the statement that parallel lines are equidistant from one another is equivalent to Postulate 5.

EXERCISES

The following exercises refer to the preceding discussion and are to be done on the basis E.

1. It can be shown (Ex. 2) that lines AB and CD (Fig. II, 1) have a common perpendicular and hence do not meet. Are they therefore parallel in the sense of Posidonius? Justify your answer.

2. (a) Prove that $\measuredangle ACD = \measuredangle BDC$ in Fig. II, 1. (b) Then show that the line through the midpoints of $\overline{AB}$ and $\overline{CD}$ is perpendicular to both segments.

3. If one of Posidonius's followers had assumed that there is a third point on line CD (Fig. II, 1) whose distance to line AB is equal to AC, he could then have proved Substitute (k) in Chapter I, Section 6, and hence Postulate 5. Verify this. [See Ex. 2(a).]

4. If a follower of Posidonius had assumed that three of the points on the locus h (Fig. II, 2) were collinear, show that he could have proved Postulate 5. (See Ex. 3.)

4. PTOLEMY AND PROCLUS

Most of those who tried to prove Postulate 5 did so on the basis E. Thus they retained all of Euclid's definitions, all of his assumptions except Postulate 5, and hence all of his propositions which do not depend on Postulate 5. In particular, then, they were free to use any of the first 28 propositions in Book I. In the rest of this chapter we shall consider some of their efforts.

One of the earliest attempts of this sort was made by Ptolemy, the noted astronomer, who lived in Alexandria during the second century A.D. Inasmuch as Euclid had used Postulate 5 for the first time in proving Proposition 29, Ptolemy's plan was (1) to prove this proposition without using the postulate, and (2) to deduce the postulate from the proposition. His handling of step (2) was correct and we need not consider it (see I, §6, Ex. 3).* His reasoning in step (1) was essentially as follows. Let the parallel lines g, h (Fig. II, 3) be cut by a transversal. Suppose that

$$a + b < \text{two right angles.} \qquad (1)$$

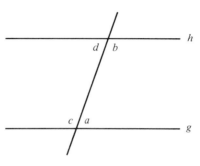

Fig. II, 3

Since g and h are as parallel on one side of the transversal as on the other, *the sum of the interior angles on one side is the same as the sum of the interior angles on the other side.* Hence

$$c + d < \text{two right angles.} \qquad (2)$$

From the inequalities (1) and (2) we then get

$$a + b + c + d < \text{four right angles.}$$

This is impossible since, by Proposition 13, $a + b + c + d$ equals four right angles. A like contradiction results if we assume that $a + b$ is greater than two right angles. Hence $a + b$ equals two right angles. The rest of Proposition 29 follows immediately.

The "as parallel as" idea used in the preceding argument is too vague to serve as mathematical justification of the italicized statement. The statement must therefore be regarded as an assumption. The reader can show (Ex. 1) that it is equivalent to Postulate 5. Thus Ptolemy proved the postulate only by making an equivalent assumption, which, moreover, is no more self-evident than the postulate.

It is to Proclus, an Athenian mathematician, philosopher, and historian of the fifth century A.D., that we owe our knowledge of Ptolemy's attempted proof. After criticizing the proof, Proclus offered his own solution of the problem. His plan was (1) to prove on the basis E that a line which meets one of two parallels also meets the other, and (2) to deduce Postulate 5 from

---

* I, §6 means Chapter I, Section 6. We shall often abbreviate cross references this way.

this proposition. Here, too, step (2) was correctly handled. The argument in step (1) runs substantially as follows. Let g, h (Fig. II, 4) be parallel lines and let another line k meet h in A. From B, a point of k situated between g and h, drop a perpendicular to h. As B recedes indefinitely far from A, its distance BC from h increases and exceeds any value, however great. In particular, BC will exceed the distance* between g and h. For some position of B, then, BC will equal the distance between g and h. When this occurs, k will meet g.

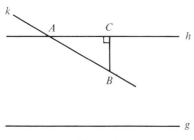

Fig. II, 4

There are a number of assumptions here which go beyond those found in Euclid. We mention only the following two: that the distance from one of two intersecting lines to the other increases beyond all bounds as we recede from their common point, and that the distance between two parallels never exceeds some finite value. Making the first of these assumptions was not a grave error insofar as Proclus's main problem was concerned, for the assumption can be proved to be a theorem on the basis E. The same cannot be said for the second assumption, which, as we saw in Chapter I, Section 6, is equivalent to Postulate 5. Thus Proclus fared no better than Ptolemy in his attempt to prove the postulate.

EXERCISES

1. Show that Ptolemy's assumption is equivalent to Postulate 5.

2. Let lines g, h have a common perpendicular line k, let A be a point of g not on k, and let B be the projection of A on h. Using only the basis E show that there is a point C on g, different from A, whose distance from h equals AB.

3. Let lines g, h have no common perpendicular, let A be a point of g, and let B be the projection of A on h. Using only the basis E show that there is no point on g other than A whose distance to h equals AB. [See §3, Ex. 2(b).]

---

* When we speak of the distance between two lines, or from one line to another, or from a point to a line, we always mean perpendicular distance.

5. SACCHERI

In the long interval from Proclus (fifth century A.D.) to the Italian logician and mathematician G. Saccheri (1667–1733), continued attempts to solve the problem were made, first by the Arabs, who had become the new leaders in geometry, and then, during the Renaissance, by Italians, Frenchmen, and Englishmen. No one obtained a genuine solution, although many thought they had. Usually the reasoning was marred by the use of some unproved statement, or unjustified procedure, equivalent to Postulate 5. Starting with Saccheri, the attempted solutions of the problem became bolder and more creative, leaving in their wake many new and valuable ideas. Some of these attempts involved the revolutionary procedure of trying to prove Postulate 5 on the basis *E* by assuming a proposition contrary to the postulate and reaching a contradiction. Saccheri seems to have been the first to follow this course.

Taking a quadrilateral *ABCD* (Fig. II, 5) in which sides $\overline{AD}$, $\overline{BC}$ are equal, and perpendicular to the base $\overline{AB}$, he proved that the angles at *C* and *D* are equal. These angles, often called the *summit angles* of the quadrilateral, are therefore either right, obtuse, or acute. To assume that they are

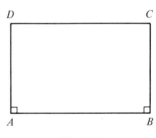

Fig. II, 5

either obtuse or acute is to deny that Postulate 5 holds, for if the latter held, the angles would of course be right angles. Saccheri's plan was therefore to assume, in turn, that they are either obtuse or acute, be led in each case to a contradiction, infer that they are therefore right angles, and go on to deduce Postulate 5 from this fact. He handled this last step correctly. Also, he reached a genuine contradiction from assuming that the summit angles are obtuse, which he called the *hypothesis of the obtuse angle*. The infinitude of straight lines played a part in reaching the contradiction since Euclid's Propositions 16, 17, and 18 were used. Before reaching the contradiction, however, Saccheri had correctly proved that·

1. $AB > CD$ in Fig. II, 5.
2. The angle-sum of a triangle always exceeds two right angles.
3. An angle inscribed in a semicircle is always obtuse.

The student may feel that these propositions, too, represent contradictions, but this is not so. They simply differ from familiar facts of Euclidean geometry inasmuch as they are obtained from a basis other than the one used by Euclid.

Next, Saccheri tried to reach a contradiction by assuming that angles C and D are acute, which he called the *hypothesis of the acute angle*. Again, because he was working with a different basis from that of Euclid, Saccheri reached unfamiliar conclusions, a great many of them this time. The following is a partial list.

1. $AB < CD$ in Fig. II, 5.
2. The angle-sum of every triangle is less than two right angles.
3. An angle inscribed in a semicircle is always acute.
4. If two lines are cut by a transversal so that the sum of the interior angles on the same side of the transversal is less than two right angles, the two lines do not necessarily meet, that is, they are sometimes parallel.
5. Through any point not on a given line there passes more than one parallel to the line.
6. Two parallels need not have a common perpendicular.
7. Two parallels are not equidistant from one another. When they have a common perpendicular, they recede from one another on each side of this perpendicular (Fig. II, 6).* When they have no common perpendicular, they

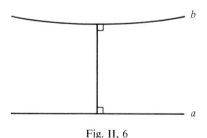

Fig. II, 6

recede from each other in one direction and are asymptotic in the other direction (Fig. II, 7).*

---

* In a diagram illustrating this property it is convenient to make at least one of the straight lines look curved. In Saccheri's proof, however, this line is no different from other lines.

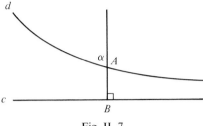

Fig. II, 7

8.   Let c, d be two parallels which are asymptotic to the right (Fig. II, 7), A any point on d, and B the projection of A on c. Then α, the angle between d and line AB, is acute, always increases as A moves to the right, and approaches a right angle when A moves without bound in that direction.

Following his proof of statement 8 Saccheri's arguments become strongly intuitive, involving the vague concept of infinitely distant points. He infers from the statement, after a long argument, that line AB approaches a limiting line g (Fig. II, 8) as A moves without bound to the right, that g is perpen-

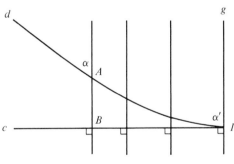

Fig. II, 8

dicular to c, that c and d meet at an infinitely distant point I on g, and that the angle α' between d and g is a right angle. The distinct lines c and d are therefore both perpendicular to g at I. Saccheri regards this as a contradiction of Euclid's Proposition 12, according to which there is a unique perpendicular to a line at each point of the line, and thus believes he has shown the hypothesis of the acute angle to be untenable. However, his reasoning involves several unsupported claims, among them the assertion that lines c and d meet. To be sure, they do get closer and closer, and intuition suggests that they meet far off, but this does not constitute mathematical proof. Had he shown by a sound argument using only the basis E that c and d meet, this would have represented a genuine contradiction since, by assumption, they are parallel.

Thus Saccheri was wrong in thinking that he had reached a contradiction. If we therefore strike out from his work on the hypothesis of the acute angle the rather small part dealing with infinitely distant elements, what remains are a considerable number of theorems, including the eight listed above, all correctly proved and consistent with one another, although clashing with familiar facts of Euclidean geometry. This clearly suggests that a new system of geometry might exist whose fundamental assumptions are the ones used by Saccheri in proving those theorems, namely, the basis E and the hypothesis of the acute angle. Going hand in hand with this suggestion is the further idea that perhaps Postulate 5 cannot be proved on the basis E, for theorem 4 in the list above clearly contradicts the postulate.

EXERCISES

1. Show that Saccheri's quadrilateral can be constructed on the basis E.

2. What was Saccheri's main objective and what was his plan for achieving it?

3. Did Saccheri correctly show that the hypothesis of the obtuse angle was untenable? the hypothesis of the acute angle? Explain.

4. State three theorems correctly deduced by Saccheri from his hypothesis of the acute angle.

Exercises 5 and 6 refer to the quadrilateral $ABCD$ (Fig. II, 5) used by Saccheri. In these exercises F and G denote the midpoints of $\overline{AB}$ and $\overline{CD}$.

5. Using only the basis E prove the following in the given order:

(a) $\angle C = \angle D$.
(b) Line FG is perpendicular to lines AB and CD.
(c) Lines AB and CD are parallel.

6. Prove the following in the given order, using only the basis E and the hypothesis of the acute angle. You may assume that this hypothesis holds for every quadrilateral containing two right angles and two equal sides situated as in quadrilateral $ABCD$. (See the remark preceding Ex. 5.)

(a) $BC > FG$.
(b) There exist parallel lines which are not equidistant [see Ex. 5(c)].

7. Show that the hypotheses of the obtuse angle and the acute angle imply, respectively, that triangles exist having angle-sums greater than two right angles and angle-sums less than two right angles.

6. LAMBERT

The publication of Saccheri's work* in 1733 attracted considerable attention and therefore could hardly have failed to affect the thinking of other investigators of the problem. A dissertation by the German mathematician G. S. Klügel in 1763, for example, carefully examines Saccheri's work, along with other celebrated attempts to prove Postulate 5, and in its conclusion expresses doubt, apparently for the first time, as to its demonstrability. Although such doubts became more and more common in the decades which followed, they also went hand in hand with further attempts to prove the postulate. Klügel's work, and hence in all likelihood Saccheri's, was known to the Swiss mathematician J. H. Lambert (1728–1777), who, in his *Theorie der Parallellinien* (1766), critically examines the question of the provability of the postulate, following this with an attempt of his own to prove it on the basis *E*.

Lambert's method is like Saccheri's and his fundamental figure is a quadrilateral *ABCD* with three right angles, say at *A*, *B*, *C*. (Such a quadrilateral is often called a *Lambert quadrilateral*, and the type used by Saccheri a *Saccheri quadrilateral*.) Under Postulate 5, angle *D* (often called the *fourth angle* of the quadrilateral) would be a right angle. Hence to assume that it is either obtuse or acute is to deny the postulate. Calling these assumptions the *hypothesis of the obtuse angle* and the *hypothesis of the acute angle*, Lambert determines their consequences, hoping to reach contradictions. He succeeds in the first case, but not in the second, well aware that he has reached no contradiction. Despite the inconclusive character of his investigations, which were not made public until 1786, they contain much of value.

It can be shown that Lambert's two hypotheses are consequences, respectively, of Saccheri's same-named hypotheses, and hence that his work, in a logical sense, is a continuation of Saccheri's. Like Saccheri he took for granted that straight lines are infinite and used this assumption in eliminating the hypothesis of the obtuse angle. After deducing from the hypothesis of the acute angle that the angle-sum of a triangle is less than two right angles, he went on to show that the sum increases when the area decreases. He did this by considering the (positive) difference between the angle-sum and two right angles, which is called the *defect* of the triangle, and proving that the defect is proportional to the area, in symbols, $D = kA$. Thus when A approaches zero, so does D, and the angle-sum approaches two right angles.

Lambert also discovered another remarkable consequence of his hypothesis of the acute angle. Just as special angles, which we name right angles, straight angles, 45° angles, and so on, can be defined and constructed, all

---

* It bears the title *Euclides ab omni naevo vindicatus* (Euclid Freed of Every Blemish).

angles of the same name being congruent, so Lambert observed that special segments, too, can be defined, constructed, and named, any two segments of the same name being congruent, and any two of different names not being congruent. This cannot be done in Euclidean geometry. Of course, we can construct three-inch segments, two-foot segments, and so on, but such constructions are physical and have no place in Euclidean geometry. Even the unit segments so often used in Euclidean geometry are not special segments as this term is used above, for unit segments are not all congruent inasmuch as any segment may be called a unit segment. For example, we often regard the legs of an isosceles right triangle as unit segments, and in dealing with a circle we frequently take its radius to be a unit segment.

When one of Lambert's special segments is used as a unit segment, it is called an *absolute unit of length*. By contrast, a unit segment in Euclidean geometry is called a *relative unit of length*. Although there are no absolute units of length in Euclidean geometry, there are absolute units of angle, as our discussion has made clear. A right angle is an example of such a unit, and so is a degree.

EXERCISES

1. Using only the basis E, show how to construct a quadrilateral with three right angles.

2. Using only the basis E, show that any Saccheri quadrilateral can be subdivided so as to give two congruent Lambert quadrilaterals. Hence show that each of Lambert's hypotheses implies the same-named Saccheri hypothesis.

3. Using only the basis E, show that from any given Lambert quadrilateral a certain Saccheri quadrilateral can be determined. Hence show that each Saccheri hypothesis implies the same-named Lambert hypothesis.

4. Using only the basis E and Lambert's hypothesis of the acute angle, show that if a triangle is subdivided so as to give two triangles, its defect is the sum of theirs, and hence its angle-sum is less than each of theirs.

7. LEGENDRE

Perhaps the last notable attempts to prove Postulate 5 were those of A. M. Legendre (1752–1833), the French mathematician. His investigations on the subject, given in the various editions of his *Eléments de Géométrie* (1794–1823), were brought together in a publication which appeared in 1833, the year of his death. Although his work gained a wide circle of readers, this was due primarily to the clarity and simplicity of his writing rather than to the discovery of new results, of which he offered very little. Like Saccheri and Lambert he tried to deduce Postulate 5 from the basis E by use of a *reductio ad absurdum* argument. Knowing that he could succeed if he could show that the angle-sum of a triangle is two right angles [see Substitute (e) in I, §6], he tried to rule out the possibility that this sum is more than two right angles or less than two right angles, situations which correspond, respectively, to his predecessors' hypotheses of the obtuse angle and the acute angle. He correctly showed that the sum cannot be more than two right angles, again using the infinitude of straight lines to do this. In one of his attempts to prove that the sum cannot be less than two right angles, he proceeded essentially as follows:

Assume that ABC (Fig. II, 9) is a triangle whose angle-sum is less than

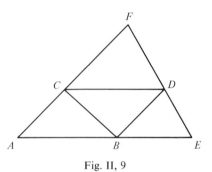

Fig. II, 9

two right angles. Using degree measure for convenience, we can then say that the angle-sum is $180° - \alpha$, where α is the defect of the triangle in degrees. Construct triangle BCD by drawing $\angle BCD$ equal to $\angle ABC$, taking CD equal to AB, and joining B to D. This triangle, being congruent to the original one, has the same angle-sum $180° - \alpha$. Through D now draw a line which meets lines AB and AC, say in E and F, forming the new triangles BDE and CDF. Since the angle-sums of these triangles are not more than two right angles, or $180°$, the sum of all the angles of the four triangles thus far considered is not more than $720° - 2\alpha$. Of these angles, those with vertex B add up to $180°$, and likewise those with vertex C, and those with vertex D. Hence the angle-sum of triangle AEF is not more than $180° - 2\alpha$. The defect of this triangle is therefore at least 2α, whereas that of the original triangle ABC is α. Thus we have a construction for doubling (at least) the defect of a triangle. Applied to triangle AEF, this construction will yield another triangle whose defect is at least 4α, and by applying the construction sufficiently many times we can obtain a triangle whose defect is arbitrarily great. But, by its nature, the defect of a triangle must be less than $180°$. Thus a contradiction has been reached and the angle-sum of a triangle cannot be less than two right angles.

There is nothing logically wrong with this proof, but it rests on more than the basis E. In taking the liberty of drawing a line through D that meets lines AB and AC, Legendre was, in effect, making the assumption that through any point within an angle a line can be drawn which meets both sides of the angle. This assumption is equivalent to Postulate 5, as was noted in Chapter I, Section 6, where it is listed as Substitute (j). Since we never proved this equivalence, let us do so now.

First, from Legendre's work we see that Postulate 5 can be proved if we use statement (j) and the basis E. Conversely, we must show that statement (j) can be proved if we use Postulate 5 and the basis E. Consider any angle with vertex A and sides g, h, and any point D within this angle (Fig. II, 10). Draw

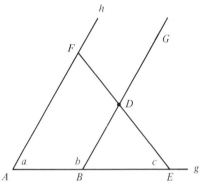

Fig. II, 10

line DG parallel to h (Prop. 31). Since h meets g, so does line DG, say in B. Take any point E on g to the right of B. Draw line DE. Since $a + b$ equals two right angles (Prop. 29) and $c < b$, it follows that $a + c$ is less than two right angles. According to Postulate 5, then, line DE must meet side h in a point F. This proves statement (j).

EXERCISES

1. Using only the basis E, show that if the angle-sum of a triangle is always less than 180°, then the fourth angle in a Lambert quadrilateral is always acute.

2. It can be shown on the basis E that if the angle-sum is less than 180° in one triangle, then it is less than 180° in all triangles. Using this fact, verify that in Legendre's proof the defect of triangle AEF (Fig. II, 9) is actually more than 2α.

8. THE DISCOVERY OF NON-EUCLIDEAN GEOMETRY

Well written as it was, Legendre's 1833 publication viewed from our time was out of date even before it appeared, so greatly had some of his contemporaries progressed with the problem of Postulate 5. Not only had there been a rapidly growing belief that the postulate could not be proved without using some equivalent statement, but, more important for the further development of geometry, the conviction had formed that there was no flaw in the system of propositions deducible from Saccheri's hypothesis of the acute angle, and hence that this system constituted a new geometry. The men who conceived and developed this new geometry did their initial work on it approximately in the period 1800 to 1830. Since they worked for the most part independently of one another and lived in different countries, there is considerable variety in their approach to the new system and in the names they gave it. Among these names we find anti-Euclidean geometry, astral geometry, non-Euclidean geometry, logarithmic-spherical geometry, imaginary geometry, and pangeometry. Of these, only the name non-Euclidean geometry is still used today.

The rapidity with which the old epoch was dying and the new one was being born can be seen from the fact that the discoverers of non-Euclidean geometry, namely, Bolyai, Gauss, Lobachevsky, Schweikart, Taurinus, and Wachter, had, except perhaps for Schweikart, all tried without success to prove Postulate 5. C. F. Gauss (1777–1855) had the distinction of being the first of these men to glimpse and develop the new geometry, as we learn from his correspondence and private papers, but he refrained from publishing his results lest their unpopularity damage his great reputation. F. K. Schweikart (1780–1857) and F. L. Wachter (1792–1817) likewise published nothing. F. A. Taurinus (1794–1874) was the first to publish his results, which were substantial, but while accepting the new geometry as logically tenable he rejected it on physical grounds, believing Euclidean geometry to be the only possible system applicable to space. J. Bolyai (1802–1867) and N. I. Lobachevsky (1793–1856) are usually accorded top honors with Gauss as the discoverers of non-Euclidean geometry, for they not only published extensive developments of the subject, believing it to be logically sound, but were convinced that it was as applicable to the physical world as Euclidean geometry. Since a lengthy and systematic discussion of the new geometry is given in the chapters which follow, we refrain from describing the separate contributions of its discoverers and their individual thinking. This, of course, need not prevent the student from making his own investigation of that fascinating subject.

Throughout the nineteenth century non-Euclidean geometry remained a subject of great interest and study. Although its discoverers had strongly believed that it involved no logical contradictions and that Postulate 5 could not be proved on the basis E, they offered no formal demonstration to support this view. In the latter part of the nineteenth century the impossibility of proving Postulate 5 on the basis E was demonstrated. Also, it was shown that if there was a logical contradiction within non-Euclidean geometry, then there would have to be one in Euclidean geometry. Thus, if Euclidean geometry is consistent, so is non-Euclidean geometry.

Another important discovery made in the latter part of the preceding century, mainly as a result of the thinking of G. F. B. Riemann (1826–1866), was that, besides the geometry of Bolyai, Gauss, and Lobachevsky, there is also a type of non-Euclidean geometry in which straight lines are not assumed to be infinite and in which there are no parallel lines. The simultaneous occurrence of these two properties will not surprise us if we recall that the existence of parallel lines in Euclid's system is an indirect consequence of his assumption that straight lines are infinite. If we recall, further, that this same assumption enabled Saccheri to reject his hypothesis of the obtuse angle as logically untenable, then we will again not be surprised to learn that in this second type of non-Euclidean geometry the summit angles of a Saccheri quadrilateral are obtuse and the angle-sum of a triangle exceeds two right angles.

For reasons we will not go into because of their unimportance for our purpose, it is now customary to call the non-Euclidean geometry of Bolyai, Gauss, and Lobachevsky *hyperbolic geometry* and the type of non-Euclidean geometry suggested by Riemann's thinking *elliptic geometry*.

HYPERBOLIC GEOMETRY

III

Parallels With
a Common Perpendicular

1. INTRODUCTION

The non-Euclidean geometry of Bolyai, Gauss, and Lobachevsky, or hyperbolic geometry, is the system of geometry which rests logically on the basis E and an assumption contradicting Euclid's parallel postulate. There are many possible forms which that assumption, or *hyperbolic parallel postulate*, can take and in Section 4 we shall make our choice. Prior to that we shall examine the basis E more closely and call attention to the theorems of hyperbolic geometry which are deducible from it without the aid of the parallel postulate. Those theorems, of course, also belong to Euclidean geometry.

In hyperbolic geometry, as Saccheri's work showed, there are two kinds of parallel lines, those with a common perpendicular and those without a common perpendicular. The first kind being more familiar and easier to deal with, we shall consider it and the related parts of hyperbolic geometry first, reserving the present chapter for that purpose. In the next chapter the second kind of parallel lines is considered.

2. THE BASIS E

We have been designating as basis E the set of all Euclid's definitions and assumptions, stated and unstated, except his parallel postulate. Of the unstated assumptions only one has been mentioned thus far, the one in which Euclid took for granted that a straight line is infinite, and now we wish to call attention to others that will be needed in developing hyperbolic geometry. Before stating them formally let us try to throw some light on their general character by referring directly to Euclid's work.

In proving Proposition 1 he draws two circles, one with center A and passing through B, the other with center B and passing through A, and takes for granted that the circles meet. There is no axiom or postulate to justify this and, of course, no previous proposition. It therefore represents an unstated assumption. Then, in the proof of Proposition 10, after taking a line AB and a point C not on it, he says "let a point D be taken at random on the other side of line AB." This involves the unstated assumption that there are such things as sides of a line and that there are two of them. In proving Proposition 21 he considers a triangle ABC and lets a line through B and a line through C meet in a point D "within the triangle." Since he never defines the term "within," this proof involves the unstated assumption that a triangle has an inside and, presumably, an outside. Also, later in the proof of the same proposition he takes for granted that line BD meets side AC in a point E.

These few examples, chosen among many, well illustrate the fact that Euclid's unstated assumptions involve geometric properties that are intuitively very acceptable. So much so, that the discoverers of non-Euclidean geometry had no quarrel with them. And neither have we, except that in keeping with current standards of rigor we shall make our recognition of the unstated assumptions explicit by listing those which are used in our work. Although we are regarding the listed properties as assumptions, in modern rigorous presentations of Euclidean and hyperbolic geometry almost all of them would be proved.

In our list, whenever possible we have placed together, under separate headings, properties which are of a similar nature. Actually, some of the listed properties follow from others, but this need not concern us. Two familiar propositions on congruent triangles have been included because Euclid's proof of them uses the unacceptable method of superposition, that is, of applying one triangle to another as if they were physical objects. Also, we have included a number of useful properties which, strictly, cannot be said to have been taken for granted by Euclid since they occur nowhere in his work. However, these properties are either so closely related to things he took for granted or are such fundamental facts of Euclidean geometry as we know the subject today, that they may properly be regarded as part of the basis E. In the interest of clarity we have redefined many of the familiar terms which appear in the list.*

* We suggest the possibility that the list could be treated as reference material, to be consulted when needed.

Separation Properties

1. Any line g separates the rest of the plane into two regions called *half-planes* (or *sides* of g). The segment joining two points in different half-planes meets g; the segment joining two points in the same half-plane does not meet g, and hence all of its points are in that half-plane. If A is any point on g, and B is any point not on g, then all points between A and B lie in the same half-plane as B.

2. Two intersecting lines separate the rest of the plane into four regions. If points A, B belong to the same region, so does each point of $\overline{AB}$; if A, B belong to different regions, then $\overline{AB}$ meets at least one of the lines. Two parallel lines separate the rest of the plane into three regions. If A is any point on one line, and B is any point on the other, then each point P between A and B is said to lie *between the two parallels*. The set of all such points P constitutes one of the three specified regions, which is therefore said to lie between the other two regions or between the two parallels.

3. Any point A on a line separates the rest of the line into two parts, called *rays* (or *half-lines*) with *endpoint A*. The point A is on neither ray, but is between each point of one ray and each point of the other. A ray with endpoint A which contains a point B is called the *ray AB*. The figure consisting of two rays AB, AC is called an *angle* and denoted by $\measuredangle BAC$ or $\measuredangle CAB$; the rays are called the *sides* of the angle and A is called its *vertex*. An angle with collinear rays is called a *straight angle*.

4. An angle (together with its vertex) separates the rest of the plane into two regions. If an angle is not a straight angle, these regions are called the *interior* and the *exterior* of the angle. The interior (but not the exterior) has the property that if two points belong to it, so do all points between them. If two points are on different sides of an angle other than a straight angle, each point between them lies in the interior of the angle.

5. A triangle separates the points of the plane not on it into two regions, called the *interior* and the *exterior* of the triangle. An interior point is characterized by the fact that it lies between two points on different sides of the triangle. The segment joining an interior point and an exterior point contains a point of the triangle.

Subdivision Properties

6. If a ray is situated so that its endpoint coincides with the vertex of an angle and the ray contains an interior point of the angle, then all points of the ray are interior to the angle. We shall say that such a ray (and also the line containing it) *subdivides* the angle. The ray and the line will be called *subdividers* of the angle. If a ray OG subdivides an angle $\measuredangle POQ$, then $\measuredangle POG + \measuredangle GOQ = \measuredangle POQ$ (Fig. III, 1). If another ray OH also subdivides

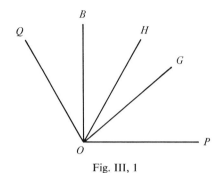

Fig. III, 1

$\angle POQ$, and $\angle POG$ is less than $\angle POH$, we shall say that ray *OG precedes* ray *OH* (also that line *OG* precedes line *OH*).

7.   A line through the intersection of two others subdivides just one pair of the vertical angles formed by the two lines.

8.   A line joining a vertex of a triangle to an inner point of the opposite side subdivides the angle at that vertex.

9.   A line joining a vertex of a quadrilateral to an inner point of a non-adjacent side subdivides the angle at that vertex. (By a quadrilateral we shall always mean a *convex quadrilateral*, which is the type dealt with in high school geometry. If $\overline{AB}$ is any side of a convex quadrilateral, then, by definition of such a quadrilateral, all the points of the other sides lie in the same one of the two half-planes determined by line *AB*. The diagonals of a convex quadrilateral always meet.)

10.   A line joining two nonadjacent vertices of a quadrilateral subdivides the angles at those vertices.

Intersection Properties

11.   A line which subdivides an angle of a triangle intersects the opposite side, that is, meets it in an inner point.

12.   A line that intersects a side of a triangle, but does not go through a vertex, also intersects exactly one other side. (This assumption is known as the *Axiom of Pasch.**)

Miscellaneous Properties

13.   Two triangles are congruent if their sides are equal, respectively, or if two sides and the included angle of one triangle are equal to two sides and the included angle of the other.

---

* M. Pasch was a nineteenth-century mathematician whose work helped to strengthen the logical foundation of Euclidean and hyperbolic geometry.

14.   If the lines subdividing an angle are formed into two sets S, T such that each line in S precedes each line in T, then either S has a last line (a line preceded by every other line in S) or T has a first line (a line preceding every other line in T). This special subdivider is called the *boundary* between S and T.

[For example, if the lines that subdivide $\not\!\prec POQ$ (Fig. III, 1) are formed into a set S whose members make acute angles with line OP and a set T whose members do not make such angles, each member of S precedes each member of T. According to Assumption 14, then, there is a boundary between S and T. Clearly, the subdivider OB such that $\not\!\prec POB = 90°$ is the boundary. It belongs to T and precedes every other member of T.]

15.   If A, B are fixed points and P is a variable point that approaches B, then the distance AP approaches the distance AB and line AP approaches line AB. If P is a variable point on side $\overline{AB}$ of any triangle ABC and it approaches A, then $\not\!\prec CPB$ approaches $\not\!\prec CAB$ and $\not\!\prec ACP$ becomes arbitrarily small. As a point traverses a line, its distance to any other line either remains constant or varies continuously.

3.   THE INITIAL THEOREMS OF HYPERBOLIC GEOMETRY

Since Propositions 1 through 28 in Book I of Euclid's *Elements* are proved by using only the basis E, they also belong to hyperbolic geometry. We shall therefore call them Theorems 1 through 28 in our development of that subject.* Thus, expressed briefly, Theorem 8 states that two triangles are congruent if the three sides of one are equal, respectively, to the three sides of the other and Theorem 12 states that a perpendicular can be dropped from a point to a line. Theorem 16, we recall, deals with the exterior angles of a triangle and implies that parallel lines exist, and Theorems 27 and 28 show how to construct them.

We shall now state as Theorems 29 through 33 several additional facts that rest only on the basis E. They are not included in the *Elements*, but were discovered in attempts to prove Postulate 5. Before proving them, let us agree that in a Saccheri quadrilateral $ABCD$ (Fig. III, 2), with right angles at A and B, $\overline{AB}$ will be called the *base*, $\overline{CD}$ the *summit*, and $\overline{AD}$, $\overline{BC}$ the *arms*.

Theorem 29.   *The summit angles of a Saccheri quadrilateral are equal.*

Proof.   Consider the Saccheri quadrilateral $ABCD$ (Fig. III, 2). Triangles ABC, BAD are congruent by side–angle–side (Theo. 4). Hence $AC = BD$.

---

* Since Euclid proved Theorems 4 and 8 by superposition, we are regarding the facts they state as assumptions (see §2, Property 13).

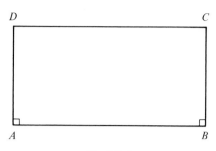

Fig. III, 2

Triangles ADC, BCD are congruent by side–side–side (Theo. 8). Therefore $\angle ADC = \angle BCD$.

Theorem 30. *The line joining the midpoints of the base and summit of a Saccheri quadrilateral is perpendicular to each. The base and summit therefore lie on parallel lines having a common perpendicular.*

Proof. Add to Fig. III, 2 the midpoints E, F of $\overline{AB}$, $\overline{CD}$ and obtain Fig. III, 3 (Theo. 10). Triangles AED, BEC are congruent by side–angle–side.

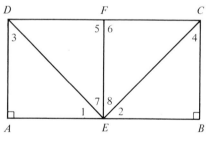

Fig. III, 3

Hence $DE = CE$, $\angle 1 = \angle 2$, and $\angle 3 = \angle 4$. Triangles CEF, DEF are congruent by side–side–side. Hence $\angle 5 = \angle 6$, so that each is a right angle. Also, $\angle 7 = \angle 8$, so that $\angle 1 + \angle 7 = \angle 2 + \angle 8 =$ a right angle. Thus, line EF is perpendicular to lines AB and CD. Lines AB and CD are therefore parallel (Theo. 27) and have a common perpendicular.

Theorem 31. *The angle-sum of a triangle does not exceed two right angles, or 180°.*

Proof. Assume that there is a triangle ABC (Fig. III, 4) whose angle-sum is $180° + \alpha$, where α is a positive number of degrees. Let D be the

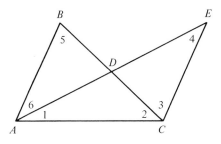

Fig. III, 4

midpoint of $\overline{BC}$. Take E on line AD so that $AD = DE$. Triangles BDA, CDE are then congruent by side–angle–side. We shall now compare triangles AEC, ABC and show that (a) they have the same angle-sum, and (b) at least one of the angles CAE, CEA does not exceed $\frac{1}{2} \measuredangle BAC$. To prove (a) we have

$$\text{angle-sum of triangle } AEC = \measuredangle 1 + \measuredangle 2 + \measuredangle 3 + \measuredangle 4$$
$$= \measuredangle 1 + \measuredangle 2 + \measuredangle 5 + \measuredangle 6$$
$$= \text{angle-sum of triangle } ABC.$$

To prove (b) we note that

$$\measuredangle 1 + \measuredangle 6 = \measuredangle BAC$$

and reason as follows. If the angles on the left are equal, then each equals $\frac{1}{2} \measuredangle BAC$; if they are unequal, then one of them is less than $\frac{1}{2} \measuredangle BAC$. Combining both cases, we can say that at least one of the angles on the left does not exceed $\frac{1}{2} \measuredangle BAC$. Since

$$\measuredangle 1 = \measuredangle CAE \qquad \text{and} \qquad \measuredangle 6 = \measuredangle CEA,$$

statement (b) is proved. Let us suppose that the "one" mentioned in (b) is $\measuredangle CAE$. Then

$$\measuredangle CAE \le \frac{1}{2} \measuredangle BAC. \tag{1}$$

If we now apply the same process to triangle AEC as was applied to triangle ABC (that is, taking the midpoint of $\overline{EC}$, etc.), we shall obtain a triangle AFC with the same angle-sum $180° + \alpha$ and in which at least one of the angles CAF, CFA does not exceed $\frac{1}{2} \measuredangle CAE$. If that one is $\measuredangle CAF$, then

$$\measuredangle CAF \le \frac{1}{2} \measuredangle CAE. \tag{2}$$

Combining (1) and (2) gives

$$\measuredangle CAF \le \frac{1}{2^2} \measuredangle BAC.$$

If the process is applied n times, we obtain a triangle PQR with the angle-sum $180° + \alpha$ and containing an angle R such that

$$\angle R \le \frac{1}{2^n} \angle BAC. \tag{3}$$

The right side of (3) can be made arbitrarily small by taking n sufficiently large. In particular, it can be made less than α. When this is done we can write

$$180° + \alpha = \angle P + \angle Q + \angle R < \angle P + \angle Q + \alpha,$$

or

$$180° < \angle P + \angle Q.$$

This contradicts that the sum of any two angles of a triangle is less than two right angles, or $180°$ (Theo. 17). Hence Theorem 31 is proved.

Theorem 32.   *The summit angles of a Saccheri quadrilateral are not obtuse.*

Proof.   Consider the Saccheri quadrilateral $ABCD$ in Fig. III, 2. If the summit angles were obtuse, the angle-sum of the quadrilateral would exceed $360°$. The angle-sum of triangle ABC plus that of triangle ACD would then exceed $360°$.* At least one of these triangle angle-sums would therefore have to exceed $180°$. This contradicts Theorem 31. Hence the summit angles are not obtuse.

In a quadrilateral $ABCD$ (Fig. III, 5) with right angles at A and B the sides $\overline{AD}$, $\overline{BC}$ may, of course, be unequal. To deal with such quadrilaterals, which are important, let us generalize the terminology used for Saccheri

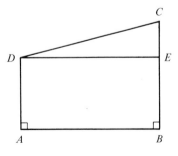

Fig. III, 5

* This conclusion is reached with the aid of Section 2, Property 10.

quadrilaterals. By a *base* of a quadrilateral we shall mean a side which is included by two right angles. The two sides adjoining a base and the two remaining angles will be called the *arms* and *summit angles* relative to the base. Thus, in Fig. III, 5, $\overline{AB}$ is a base and, relative to it, $\overline{BC}$, $\overline{AD}$ are the arms, and $\angle C$, $\angle D$ the summit angles. We shall speak of $\overline{AD}$ and $\angle C$ as being *opposite* each other, and likewise for $\overline{BC}$ and $\angle D$.

Theorem 33. *In a quadrilateral with a base, if the arms relative to the base are unequal, so are the summit angles, and conversely, the greater summit angle always lying opposite the greater arm.*

Proof. Let $ABCD$ (Fig. III, 5) be a quadrilateral with right angles at A, B, and hence with base $\overline{AB}$. Suppose $BC > AD$. On $\overline{BC}$ take point E so that $BE = AD$. Then $ABED$ is a Saccheri quadrilateral and we have

$$\angle ADE = \angle BED. \qquad (1)$$

Since line DE subdivides $\angle ADC$ of quadrilateral $ABCD$ (§2, Property 9),

$$\angle ADC > \angle ADE. \qquad (2)$$

Also, since $\angle BED$ is an exterior angle of triangle CED, we have

$$\angle BED > \angle BCD. \qquad (3)$$

On combining (1), (2), and (3) we obtain

$$\angle ADC > \angle BCD.$$

Thus the summit angles are unequal, the greater one lying opposite the greater arm. We can proceed similarly if $AD > BC$.

To prove the converse part of the theorem, assume that $\angle ADC > \angle BCD$. The arms cannot be equal since this would imply that the summit angles are equal (Theo. 29). Hence either $AD > BC$ or $BC > AD$. The first of these inequalities cannot hold, for it would imply that $\angle BCD > \angle ADC$ by the first part of the proof and this contradicts our assumption. We therefore conclude that $BC > AD$. Thus the arms are unequal, the greater one lying opposite the greater summit angle. We can proceed similarly if $\angle BCD > \angle ADC$.

EXERCISES

1. Show how the proof of Theorem 31 would have proceeded had we supposed that $\angle CEA$ does not exceed $\frac{1}{2} \angle BAC$.

2. The infinite extent of straight lines was used in the proof of Theorem 31 without explicit mention. At what step did this occur?

3. What can be inferred from Theorem 31 concerning the fourth angle of a Lambert quadrilateral? Prove it.

4. Prove that the angle-sum of a quadrilateral does not exceed 360°.

5. Show that if a quadrilateral with a base has equal summit angles relative to the base, it will have equal arms relative to the base and hence be a Saccheri quadrilateral.

6. Prove that two quadrilaterals are congruent if a side, angle, side, angle, side of one are equal to the corresponding parts of the other, the five parts in each case being consecutive on the quadrilateral.

7. Prove that two Lambert quadrilaterals are congruent if the fourth angle and the adjacent sides in one quadrilateral are equal to the corresponding parts in the other.*

4. THE HYPERBOLIC PARALLEL POSTULATE

There are many choices for the statement to serve as the hyperbolic parallel postulate. Most directly, it could be a straightforward negation of Euclid's parallel postulate, namely: Straight lines g, h (Fig. III, 6) exist which

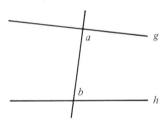

Fig. III, 6

have a transversal such that $a + b$ is less than two right angles and which do not meet. The discoverers of hyperbolic geometry did not use this, preferring to assume that through a point not on a given line there passes more than one line not meeting the given line. This assumption, which negates Playfair's Axiom, and hence Euclid's parallel postulate, has been commonly used in textbooks on the subject. However, the assumption we shall use is Saccheri's hypothesis of the acute angle, for it will enable us to start out more simply and proceed more rapidly than is possible under the traditional approach.

---

* The congruence can also be proved by assuming less. See Section 8, Exercise 8.

Before stating the assumption formally let us note that, in view of Theorems 29 and 32, there are two possibilities for the summit angles of a Saccheri quadrilateral: either they are right angles or they are acute angles. Nothing further about the size of the summit angles can be proved on the basis *E*. If now we were to adopt the additional assumption that they are right angles, this would be equivalent to adopting Euclid's Postulate 5 and thus obtaining Euclidean geometry. If, on the other hand, we assumed that the angles are acute, we would be negating Postulate 5. The latter course being the one we wish to take, we now formally adopt the following assumption:

Hyperbolic Parallel Postulate. *The summit angles of a Saccheri quadrilateral are acute.*

It might seem, at first, that this assumption is less in accord with experience than the assumption that the summit angles are right angles, but a little reflection will show that this is not necessarily the case. Ordinarily, to be sure, we do regard the summit angles as being right angles, but this is mainly because we have been conditioned to think in a Euclidean manner. The angles *appear* to be right angles and so, influenced by our study of Euclidean geometry, we say they *are* right angles. What if the angles are actually very slightly less than 90°, but so close that the difference cannot be detected by measurement? This would be consistent with the hyperbolic parallel postulate, which merely says that the angles are acute, without specifying their actual size.

5. IMMEDIATE CONSEQUENCES OF THE POSTULATE

Our hyperbolic parallel postulate, being logically equivalent to the statement negating Postulate 5, is also logically equivalent to each statement negating a substitute for Postulate 5. Hence, if we take any substitute and form the statement negating it, that statement will express a fact of hyperbolic geometry. Consider, for example, Substitute (f) in Chapter I, Section 6, which we reword for convenience:

(f)   It is always true that parallel lines are equidistant from one another.

The statement negating this is

(f̄)   It is not always true that parallel lines are equidistant from one another.

Or, in other words,

(f̄)   Parallel lines exist which are not equidistant from one another.

This is a fact of hyperbolic geometry.

To clarify the logic by which we know that $(\bar{f})$ is a fact of hyperbolic geometry, let us prove it. Assume $(\bar{f})$ is not a fact of hyperbolic geometry. Then (f), its negation, is a fact of hyperbolic geometry. Also, we know that the basis E holds in hyperbolic geometry. From (f) and the basis E one can deduce Postulate 5 since (f) is a substitute for Postulate 5. Hence Postulate 5 holds in hyperbolic geometry. In turn, from E and Postulate 5 we can prove that the summit angles of a Saccheri quadrilateral are right angles. This contradicts the hyperbolic parallel postulate. Hence statement $(\bar{f})$ is a fact of hyperbolic geometry.

Similarly, by negating Substitute (a) in Chapter I, Section 6, after adding "always" for emphasis, we obtain the statement: It is not true that through a point not on a given line there always passes not more than one parallel to the line. This is equivalent to saying: A line and a point not on it exist such that more than one parallel to the line passes through the point. This is a fact of hyperbolic geometry. To give one more example, let us note that by negating Substitute (h) in Chapter I, Section 6, which states that similar triangles exist which are not congruent, we obtain the statement: Similar triangles are always congruent. This is a fact of hyperbolic geometry.

This procedure of negating substitutes for Postulate 5 is obviously useful in quickly providing us with salient facts of hyperbolic geometry. Sometimes the facts thus obtained are complete and tell us everything we could wish to know about the corresponding situations. An example of this is the statement that similar triangles are always congruent. Usually, however, the facts only give us some basic and important minimum of knowledge, leaving many questions unanswered. An example of this is the fact, noted above, that parallel lines exist which are not equidistant. How often can this occur? And when it occurs, just how does the distance from one line to the other vary? Then there is the fact that a line and a point not on it exist such that more than one parallel to the line passes through the point. How often does this actually happen? When it happens, how many parallels can there be? To answer questions like these, and in fact to develop most of the content of hyperbolic geometry, calls for the same kind of systematic study as that by which Euclidean geometry is learned. This study, moreover, will also provide us with direct geometric proofs of those facts which can be learned by negating substitutes for Postulate 5.

EXERCISES*

1. By negating a substitute for Postulate 5 determine a fact of hyperbolic geometry concerning each of the following:

---

* We take for granted in these exercises that all the substitutes for Postulate 5 listed in Chapter I, Section 6, have been proved.

(a)   a line which meets one of two parallels;

(b)   the alternate interior angles formed when two parallels are met by a transversal;

(c)   the angle-sum of a triangle;

(d)   the number of common perpendiculars of two parallel lines;

(e)   two lines which are parallel to the same line;

(f)   the number of circles passing through three noncollinear points.

2.   Prove that the angle-sum of a triangle in hyperbolic geometry is always less than 180°.

3.   Prove that noncongruent similar triangles do not exist in hyperbolic geometry. (A detailed proof is wanted, like the one used above to show that parallel lines exist which are not equidistant.)

4.   Prove that the statement negating any substitute for Postulate 5 is a fact of hyperbolic geometry.

5.   Prove that no two lines in hyperbolic geometry are equidistant from one another by showing that the distance from one line to another cannot have the same value in more than two places.

6.   The set of points which are at the same distance from a given line and lie on the same side of it is called an *equidistant curve*, and the line is called the *base line* of the curve. Prove that in hyperbolic geometry no three points of an equidistant curve are collinear.

7.   If the vertices of a triangle in hyperbolic geometry lie on an equidistant curve (see Ex. 6), show that the perpendicular bisectors of the sides are parallel, with a common perpendicular.

6. FURTHER PROPERTIES OF QUADRILATERALS

In Section 3, before introducing the hyperbolic parallel postulate, we stated and proved some properties of quadrilaterals. Those properties, resting only on the basis E, hold in Euclidean as well as in hyperbolic geometry. Now, using the postulate, we shall prove additional properties which hold only in hyperbolic geometry.

Theorem 34.   *In a Saccheri quadrilateral the summit is longer than the base and the segment joining their midpoints is shorter than each arm.*

Proof.   Consider a Saccheri quadrilateral $ABCD$ with right angles at A, B (Fig. III, 7). If E and F are the midpoints of the base and summit, then $\overline{EF}$ is perpendicular to the base and summit (§3, Theo. 30). Since the angles

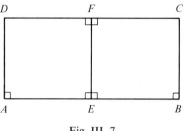

Fig. III, 7

at C and D are acute, $AD > EF$ in quadrilateral $AEFD$ and $BC > EF$ in quadrilateral $EBCF$ (§3, Theo. 33). This proves the second part of the theorem. On noting that these quadrilaterals can also be viewed so that $\overline{AE}$, $\overline{DF}$ and $\overline{EB}$, $\overline{FC}$ are arms, we again use Theorem 33, obtaining

$$DF > AE \qquad \text{and} \qquad FC > EB.$$

Combining these inequalities gives

$$DF + FC > AE + EB,$$

or

$$DC > AB,$$

which completes the proof of the theorem.

Lines AB, CD in the preceding discussion are parallel since they have a common perpendicular, and AD, EF are two distances from line CD to line AB that are unequal. Thus we have proved by a direct geometrical procedure that parallel lines exist which are not equidistant from one another. In Section 5 we reached the same conclusion by negating a substitute for Postulate 5.

The quadrilaterals $AEFD$, $EBCF$ considered in the proof of Theorem 34 are Lambert quadrilaterals and were obtained by subdividing the Saccheri quadrilateral $ABCD$. The facts observed concerning their sides and fourth angles hold for all Lambert quadrilaterals, as we now prove.

Theorem 35. *The fourth angle in a Lambert quadrilateral is acute and each side adjacent to it is longer than the opposite side.*

Proof. Let $JKLM$ (Fig. III, 8) be a Lambert quadrilateral and let its fourth angle be at L. We first show that this angle is acute. Extend $\overline{ML}$ beyond M to a point N such that $NM = ML$. Let P be the projection of N on line JK. Triangles JMN, JML are then congruent by side–angle–side.

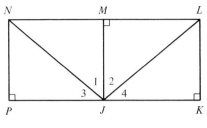

Fig. III, 8

Hence $JN = JL$ and $\not\prec 1 = \not\prec 2$. As a result, $\not\prec 3 = \not\prec 4$. Triangles JNP, JLK are then congruent by angle–angle–side (Theo. 26). It follows that $PN = KL$ and hence that $PKLN$ is a Saccheri quadrilateral. The summit angle KLN of this quadrilateral, which is acute by the hyperbolic parallel postulate, is the fourth angle of the Lambert quadrilateral $JKLM$. To show that $LK > MJ$ and $ML > JK$ one proceeds just as in the proof of Theorem 34.

EXERCISES

1. Prove that the fourth angle of a Lambert quadrilateral is acute by using an argument involving the negation of a substitute for Postulate 5.

2. Show that a quadrilateral whose opposite sides are equal is a parallelogram, but that the converse is not true. (A *parallelogram* is defined to be a quadrilateral whose opposite sides lie on parallel lines.)

3. What statement, if any, can be made about the relative lengths of summit and base in a quadrilateral known only to have two adjacent right angles? Justify your answer.

4. Show that there are no squares in hyperbolic geometry but that there are rhombuses with equal angles.

7. PARALLELS WITH A COMMON PERPENDICULAR

We have already noted that if two lines have a common perpendicular, then they are parallel (Theo. 27), as in Euclidean geometry. Such parallels are easily constructed by using a sequence of perpendiculars. If two lines have a common perpendicular, they cannot have another such perpendicular. For if they did, we would clearly have a Lambert quadrilateral in which the fourth angle is a right angle, contradicting Theorem 35.

Theorem 36. *If two parallels have a common perpendicular, then they cannot have a second one.*

Among the theorems which precede Theorem 36, only Theorems 27 and 28 (Euclid's Propositions 27 and 28) and Theorem 30 deal with parallel lines. Theorem 30 involves a special situation in which two lines are parallel and the parallels clearly have a common perpendicular. Theorems 27 and 28 state more general conditions under which two lines will be parallel. Such parallels, too, always have a common perpendicular, as we now show.

Theorem 37.   *Two lines will be parallel, with a common perpendicular, if there is a transversal which cuts the lines so as to form equal alternate interior angles or equal corresponding angles.*

Proof.   We consider only the case of alternate interior angles, for the case of corresponding angles then follows immediately. Let lines *g*, *h* be cut by a line *t* in points *A*, *B* so that $\angle 1 = \angle 2$ (Fig. III, 9). If these angles are

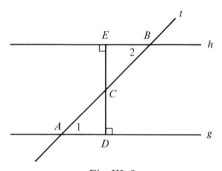

Fig. III, 9

right angles, then *t* is a common perpendicular of *g* and *h*. If they are acute angles, let *C* be the midpoint of $\overline{AB}$, and *D* the projection of *C* on *g*. On *h* take point *E* to the left of *B* so that $BE = AD$. Triangles *CAD*, *CBE* are congruent by side–angle–side. Hence $\angle BEC = \angle ADC$ and $\angle BCE = \angle ACD$. We conclude that $\angle BEC$ is a right angle and *C*, *D*, *E* are collinear. Line *DE* is therefore a common perpendicular of *g* and *h*. If $\angle 1$, $\angle 2$ are obtuse, then the other two alternate interior angles are acute and the preceding argument can be applied to them.

The proof of the following converse of Theorem 37 is left as an exercise.

Theorem 38.   *If two lines have a common perpendicular, there are transversals which cut the lines so as to form equal alternate interior angles (or equal corresponding angles). The only transversals with this property are those which go through the point on that perpendicular which is midway between the lines.*

We have already learned something about the distance between two parallel lines. First, by negating a certain substitute for Postulate 5 we saw that it is not always true that two parallels are equidistant. Then, Exercise 5 in Section 5 showed that two parallels are never equidistant. It still remains to determine precisely what the facts are. At this juncture one important theorem can be proved for parallels with a common perpendicular; another is proved later.

Theorem 39.   *The distance between two parallels with a common perpendicular is least when measured along that perpendicular. The distance from a point on either parallel to the other increases as the point recedes from the perpendicular in either direction.*

Proof.   Let *g*, *h* (Fig. III, 10) be parallels with a common perpendicular which meets them in points *E*, *F*. On *h* choose any point *B* other than *F* and let *C* be its projection on *g*. Since *ECBF* is a Lambert quadrilateral, ∡1 is acute and *BC* > *FE* (§6, Theo. 35). Thus the distance from *h* to *g* is less

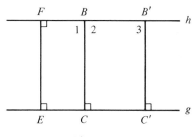

Fig. III, 10

when measured along the common perpendicular than along any other perpendicular from *h* to *g*. Similarly, by choosing *B* on *g* rather than on *h* we would see that the distance from *g* to *h* is less when measured along the common perpendicular than along any other perpendicular from *g* to *h*. This proves the first part of the theorem. Now, assuming *B* is on *h*, choose any point *B'* on *h* so that *B* is between *F* and *B'*. Let *C'* be the projection of *B'* on *g*. Then ∡3 is acute since *EC'B'F* is a Lambert quadrilateral. Also ∡2 is obtuse inasmuch as ∡1 is acute. Thus ∡2 > ∡3 in quadrilateral *BCC'B'* and we conclude that *B'C'* > *BC* (§3, Theo. 33). This completes the proof.

Theorem 39 does not tell us how great the distance between two parallels with a common perpendicular can become, nor does it say anything about the rate at which the distance increases. The rate could conceivably be very slow. Accordingly, we have no right to expect that the increase in distance can be noticed in our small diagrams when carefully drawn in the usual way, and in fact it will not be noticed. On the other hand, a diagram which obscures an important property may not be helpful as a guide in proving theorems which depend on that property. A way out of the difficulty is for us to take the liberty sometimes of drawing one or both of the parallels so that they look curved, as in Fig. III, 11.

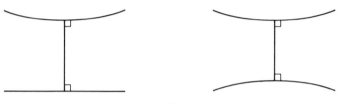

Fig. III, 11

We saw earlier, by negating a substitute for Postulate 5, that through a point not on a given line there must sometimes pass more than one parallel to the line. Now we can be more precise and prove the following.

Theorem 40.   *Given any line and any point not on it, there pass infinitely many lines through the point which are parallel to the line and have a common perpendicular with it.*

Proof.   Let g be any line, and F any point not on it (Fig. III, 12). If E is the projection of F on g, and line h is perpendicular to line EF at F, then h is one line through F which is parallel to g and has a common perpendicular

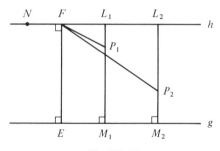

Fig. III, 12

with it. To obtain another, let L_1 be any point of h to the right of F. If M_1 is the projection of L_1 on g, then $L_1M_1 > EF$ (Theo. 39). Take P_1 on $\overline{L_1M_1}$ so that $MP_1 = EF$. Then EM_1P_1F is a Saccheri quadrilateral and lines FP_1, g are parallel, with a common perpendicular (Theo. 30). Line FP_1 is clearly distinct from h. It is therefore the second line shown to go through F which is parallel to g and has a common perpendicular with it.

Similarly, by using a different point L_2 on h to the right of F, we obtain a line FP_2 with the same properties as line FP_1, namely, distinct from h, passing through F, parallel to g, and having a common perpendicular with g. Since $EF = M_1P_1 = M_2P_2$ and the distances from one line to another cannot be equal at three different places (see §5, Ex. 5), FP_1 and FP_2 cannot be the same line. In summary, we have shown that to each point of h to the right of F there corresponds a line which goes through F, is parallel to g, and has a common perpendicular with g, and that different lines always correspond to different points. A like statement holds for the points of h to the left of F. This completes the proof.

Since $\measuredangle EFP_1$ in Fig. III, 12 is acute and $\measuredangle EFL_1$ is a right angle, line FP_1 subdivides $\measuredangle EFL_1$. From this it is clear that all the parallels FP_1, FP_2, ... corresponding to points L_1, L_2, ... to the right of F subdivide the right angle EFL, and that all the analogous parallels corresponding to points to the left of F subdivide the right angle EFN. As Fig. III, 12 correctly suggests, $\measuredangle EFP_2 < \measuredangle EFP_1$. From this we see that as a point on h is taken further from F, the acute angle which the corresponding parallel makes with line EF becomes smaller. Since there is no furthest position from F, none of *these* parallels is closest to line EF in the sense of making a smaller angle with it than do all the others. The italics emphasize that we are talking only about the special parallels obtained by our construction. Conceivably, there might be other parallels through F having common perpendiculars with g but not obtainable by that construction. We shall now show that the specified property holds for these parallels, too, should there be any.

Theorem 41. *Let g be any line, F any point not on it, and E the projection of F on g. Among all the lines through F which have a common perpendicular with g, there is none that lies closest to line EF in the sense of making a smaller angle with it than do all the others.*

Proof. Let line h be perpendicular to line EF at F (Fig. III, 13). Then h goes through F and has a common perpendicular with g, but is not the closest to line EF in view of the preceding theorem. Hence let j be any other line through F having a common perpendicular with g, and let j subdivide $\measuredangle EFL$, where L is any point of h other than F. We shall suppose, to be definite, that L is to the right of F and prove the theorem by exhibiting another line which has these same properties as j and is closer to line EF. Let A, B be the points of j, g such that $\overline{AB}$ is perpendicular to j and g. Then $\overline{AB}$ is to the right of $\overline{EF}$, for if it were to the left, say in the position $\overline{A'B'}$, $A'B'$ would

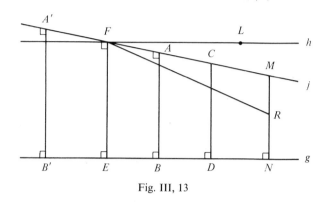

Fig. III, 13

have to exceed EF by Theorem 33 since $\measuredangle A'FE$ is obtuse, and yet be less than EF by Theorem 39. On j take C to the right of A so that $AC = AF$, and take M to the right of C. Let D, N be the projections of C, M on g. It is easily shown by using congruent triangles that $CD = EF$. Then $MN > CD$ by Theorem 39, and hence $MN > EF$. Choose R on $\overline{MN}$ so that $NR = EF$. Since $ENRF$ is a Saccheri quadrilateral, line FR has a common perpendicular with g. Also, line FR subdivides $\measuredangle EFM$ and hence is closer than j to line EF.

EXERCISES

1. Prove Theorem 38.

2. The distance between two parallels with a common perpendicular increases in the same way on one side of the perpendicular as on the other. Prove this by taking M on h to the left of F in Fig. III, 10 so that $FM = FB$ and showing that $MN = BC$, where N is the projection of M on g.

3. If two Saccheri quadrilaterals have equal arms and unequal bases, prove that the one with the greater base has smaller summit angles. (Use Ex. 2 and Theo. 39.)

4. In Fig. III, 12 prove that $\measuredangle EFP_2 < \measuredangle EFP_1$, and hence that line FP_2 is closer to line EF than is line FP_1. (Use Ex. 3.)

5. Show that two parallels with a common perpendicular are symmetrical to each other with respect to the point mentioned in Theorem 38.

6. Prove that two lines have a common perpendicular if there is a transversal such that the sum of the two interior angles on the same side of the transversal is 180°.

7. Let g be a line, F a point not on it, and E the projection of F on g. If k is a line through F which has a common perpendicular with g, prove that the minimum distance from k to g is less than or equal to EF.

8. What can be said about the angles formed when two parallels without a common perpendicular are cut by a transversal? Justify your answer.

9. Using a straightedge and compass, carefully draw a large diagram showing two parallels with a common perpendicular and measure the distance from one to the other at several places. Are your results consistent with Theorem 39? Explain.

10. If two lines have no common perpendicular, the distance from one to the other cannot have the same value at different points. Prove this.

11. Show that if a point traverses one of two lines having no common perpendicular, its distance to the other either always increases or always decreases.

8. THE ANGLE-SUM OF A TRIANGLE

Earlier, using only the basis E, we were able to show that the angle-sum of a triangle can never exceed 180° (Theo. 31). Then, on introducing the hyperbolic parallel postulate, we could have inferred immediately that no triangle can have an angle-sum of 180°. (For this is the result of negating the statement, there exists a triangle whose angle-sum is 180°, which is a substitute for Postulate 5.) We can therefore conclude that the angle-sum of a triangle is always less than 180°. Now let us offer a more familiar type of argument to arrive at this same fact. It is worthwhile to do this inasmuch as we never proved the substitute mentioned above.

Theorem 42. *The angle-sum of every triangle is less than* 180°.

Proof. We first prove the theorem for right triangles. Consider any such triangle ABC, with right angle at C (Fig. III, 14). Let BD be the line through B such that

$$\angle 1 = \angle 2. \tag{1}$$

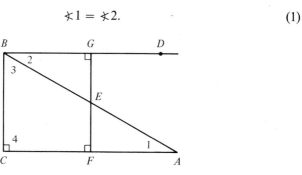

Fig. III, 14

Lines AC, BD are then parallel, with a common perpendicular (Theo. 37). Let the perpendicular meet them in F, G and let E be the midpoint of $\overline{FG}$. Then E is on line AB (Theo. 38), and midway between A and B (§7, Ex. 5). Thus the position of $\overline{FG}$ relative to triangle ABC is as shown in Fig. III, 14. We therefore obtain a Lambert quadrilateral $BCFG$ in which $\measuredangle CBG$ is acute (Theo. 35). Then, using equation (1), we have

$$\measuredangle 1 + \measuredangle 3 = \measuredangle 2 + \measuredangle 3 = \measuredangle CBG < 90°,$$

that is,

$$\measuredangle 1 + \measuredangle 3 < 90°.$$

Since $\measuredangle 4$ is a right angle, we conclude that

$$\measuredangle 1 + \measuredangle 3 + \measuredangle 4 < 180°.$$

Thus, the angle-sum of triangle ABC is less than $180°$.

Now consider any triangle PQR (Fig. III, 15) which contains no right angle. The triangle cannot contain more than one obtuse angle by Theorem 31, and hence must contain at least two acute angles. Let them be at P and Q.

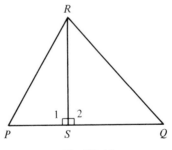

Fig. III, 15

Then S, the projection of R on line PQ, lies between P and Q, for a contradiction of Theorem 16 would result from supposing that S is outside of $\overline{PQ}$. Line RS therefore subdivides $\measuredangle PRQ$ (§2, Property 8) and we obtain two right triangles PRS and QRS. The sum of all the angles in these right triangles is less than $360°$ by the first part of this proof. Excluding the right angles $\measuredangle 1$ and $\measuredangle 2$ from this sum gives the angle-sum of triangle PQR, which is therefore less than $180°$.

Since the angle-sums of triangles thus lie between $0°$ and $180°$, one may wonder how close to these extreme values angle-sums can actually get. Later it will be possible to prove that there are triangles with angle-sums arbitrarily close to 0°. At this point we can prove

Theorem 43.   *There are triangles with angle-sums arbitrarily close to 180°.*

Proof.   Consider any triangle *ABC* (Fig. III, 16) and a variable point *D* between *A* and *B*. Let *D* approach *A*. Then ⊀*ADC* approaches ⊀*EAC*, and ⊀*ACD* becomes arbitrarily small in value (§2, Property 15). Hence the angle-sum of triangle *ADC* gets arbitrarily close to the value of ⊀*EAC* + ⊀*CAB*, which is 180°.

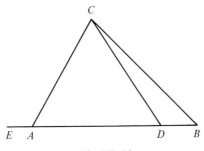

Fig. III, 16

Keeping in mind that the term "quadrilateral" always means a convex quadrilateral for us and that the line joining two nonadjacent vertices of such a quadrilateral subdivides the angles at those vertices (§2, Property 10), we can easily obtain the following corollary of Theorem 42. The proof is left as an exercise.

Theorem 44.   *The angle-sum of every quadrilateral is less than* 360°.

This theorem implies that there are no squares or rectangles in hyperbolic geometry. It also enables us to establish the following striking fact:

Theorem 45.   *Two triangles are congruent if the three angles of one are equal, respectively, to the three angles of the other.*

Proof.   Let angles *A*, *B*, *C* of triangle *ABC* (Fig. III, 17) be equal to angles *A'*, *B'*, *C'* of triangle *A'B'C'*. We shall show that the corresponding sides $\overline{AB}$, $\overline{A'B'}$ are equal and conclude that the triangles are therefore congruent by angle–side–angle (Theo. 26). Assume *AB* > *A'B'*. Take *D* on $\overline{AB}$ so that *AD* = *A'B'*. On line *AC*, on the same side of *A* as *C*, take *E* so that

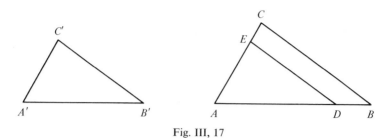

Fig. III, 17

$AE = A'C'$. Triangles $A'B'C'$ and ADE are then congruent by side–angle–side (Theo. 4). Hence $\angle ADE = \angle B$ and $\angle AED = \angle C$. It follows that E is between A and C, otherwise, as is easily verified, we would have a situation in which an exterior angle of a triangle equals an opposite interior angle of the triangle. Thus we obtain a quadrilateral $DBCE$ whose angle-sum clearly equals 360°. This contradicts Theorem 44. The assumption that $A'B' > AB$ leads to the same impasse. Hence $AB = A'B'$ and triangles ABC, $A'B'C'$ are congruent.

In Euclidean geometry, we recall, the three angles of one triangle may equal the three angles of another, respectively, without the triangles being congruent; the triangles, of course, would have proportional sides and thus be similar by definition (I, §5). Two such noncongruent similar triangles cannot exist in hyperbolic geometry according to Theorem 45. Noncongruent triangles whose sides are in proportion do exist, however. Given an equilateral triangle, for example, another with sides twice as long can be constructed just as in Euclidean geometry. Since, according to Theorem 45, the angles of the two triangles could not be equal, respectively, the triangles do not satisfy the definition of being similar.

EXERCISES

1. In the proof of Theorem 42 show in detail why S must lie between P and Q.

2. Prove Theorem 44.

3. In the proof of Theorem 45 show in detail why E is between A and C.

4. Prove Theorem 45 by negating a substitute for Postulate 5.

5. Given an equilateral triangle, construct another with sides twice as long and prove that no angle of one triangle equals an angle of the other.

6. Show that if two noncongruent Saccheri quadrilaterals have equal bases, the one with the longer arms has the smaller summit angles.

7. Prove that two Saccheri quadrilaterals are congruent* if the summit and summit angles of one are equal, respectively, to the summit and summit angles of the other.

8. Prove that two Lambert quadrilaterals are congruent* if the acute angle and an adjacent side in one are equal to the acute angle and an adjacent side in the other.

9. Prove that an exterior angle of a triangle is always greater than the sum of the two opposite interior angles.

9. THE DEFECT OF A TRIANGLE

The amount by which the angle-sum of a triangle falls short of 180° is known as the defect† of the triangle. Thus a triangle with an angle-sum of 170° has a defect of 10°. Instead of saying that there are triangles with angle-sums arbitrarily close to 180° (Theo. 43), we could say that there are triangles with arbitrarily small defects. The term "defect," however, is no mere verbal convenience, but represents an important idea, as will be seen when we study the area of a triangle.

In order to exhibit a basic fact about defects which is relevant to that study, let us now take any triangle *ABC* (Fig. III, 18) and consider the two

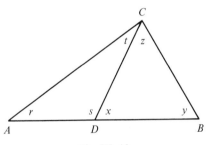

Fig. III, 18

triangles *ADC* and *BDC* formed by joining *C* to any point *D* between *A* and *B*. It will be convenient to say that the two triangles have been obtained by *subdividing* triangle *ABC*, although this terminology is rather loose.‡

---

* Two quadrilaterals are, by definition, congruent if their sides and angles are equal, respectively.

† Despite the connotation of this traditional term, there is nothing imperfect about triangles in hyperbolic geometry.

‡ Actually, it is triangle *ABC* plus its interior that is subdivided into the two triangles plus their interiors.

If d, d_1, d_2 are the defects of triangles ABC, ADC, BDC, then, as we now show,

$$d = d_1 + d_2.$$ (1)

By definition,

$$d_1 = 180° - (r + s + t) \quad \text{and} \quad d_2 = 180° - (x + y + z),$$

so that

$$\begin{aligned} d_1 + d_2 &= 360° - (r + s + t + x + y + z) \\ &= 360° - (r + t + y + z) - (s + x) \\ &= 180° - (r + t + y + z) \\ &= d. \end{aligned}$$

Stated in words, the defect of a triangle is equal to the sum of the defects of any two triangles into which it is subdivided.

Consequently, if triangle BCD is subdivided into two triangles with defects d_2' and d_2'', then

$$d_2 = d_2' + d_2''.$$

Substituting this in (1) we obtain the equation

$$d = d_1 + d_2' + d_2'',$$

relating the defect of the original triangle ABC to the defects of the three triangles into which it has been subdivided. Clearly, if any of these three triangles is subdivided into two triangles and this process of subdivision is continued further, then at each stage the defect of triangle ABC equals the sum of the defects of the triangles into which triangle ABC has been subdivided.

The method of subdivision which we have used is called *subdivision by transversals*, or *transversal subdivision*, since the segment joining a vertex of a triangle to a point on the opposite side is often called a transversal. Nontransversal subdivisions of triangle ABC into other triangles are also possible. Two of them are shown in Figs. III, 19 and III, 20 and it is easily

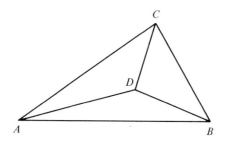

Fig. III, 19

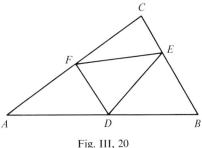

Fig. III, 20

verified that in each case the defect relation is the same as in transversal subdivisions. As might be imagined, there are many other nontransversal subdivisions of triangle ABC, but it can be proved that this same defect relation holds in all cases. The key step of the proof* consists in showing that, by the use of additional line segments, any nontransversal subdivision S of triangle ABC can be converted into a transversal subdivision S_t of triangle ABC such that each triangle resulting from S either is left unchanged by the added segments or is divided transversally by them. It then easily follows that the defect relation must hold for S inasmuch as it holds for S_t.

For example, suppose that S is the nontransversal subdivision of triangle ABC shown in Fig. III, 19. By the addition of the segment $\overline{DE}$ which is collinear with $\overline{CD}$, S is converted into the transversal subdivision S_t of triangle ABC shown in Fig. III, 21. Of the triangles CAD, BCD, ABD

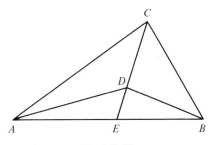

Fig. III, 21

resulting from S, the first two are unchanged by the addition of $\overline{DE}$, but the third is divided transversally by it. Since S_t is a transversal subdivision of triangle ABC, we have

$$\operatorname{def} ABC = \operatorname{def} CAD + \operatorname{def} BCD + \operatorname{def} AED + \operatorname{def} BDE, \qquad (2)$$

* D. Hilbert, *The Foundations of Geometry* (transl. by E. J. Townsend), pp. 63–65 (Open Court Publ., Chicago, Illinois, 1902).

where def is the abbreviation of defect. Also, since triangle ABD is divided transversally by $\overline{DE}$,

$$\text{def } ABD = \text{def } AED + \text{def } BDE. \tag{3}$$

Substituting (3) in (2), we get

$$\text{def } ABC = \text{def } CAD + \text{def } BCD + \text{def } ABD,$$

which shows that the defect relation holds for the nontransversal subdivision S of triangle ABC.

The relation which we have been discussing and have partially proved is described as follows.

Theorem 46.   *If a triangle is subdivided into other triangles in any manner, then the sum of the defects of those triangles equals the defect of the given triangle.*

In consequence of this theorem any triangle t resulting from a subdivision of a triangle T necessarily has a smaller defect than T. Also, since the region inside of t is part of the region inside of T, it is intuitively clear that t has a smaller area than T. Combining these two observations, we are led to conjecture that the smaller the area of a triangle, the smaller its defect, and hence the greater its angle-sum. This is correct and is proved later.

In this section we have been using degree measure for angles. As a result, the defect of a triangle was always a number of degrees. Had some other angular unit been used, such as the complete rotation, the right angle, or the *standard unit* [by definition, a *standard angular unit* equals $(180/\pi)$ degrees, or π standard units equal $180°$], the defect would have been a number of such other units. Thus, instead of saying that a triangle has a defect of $30°$, we could say that it has a defect of $\pi/6$ standard units.*

EXERCISES

1.   Show that Theorem 46 holds for the nontransversal subdivision in (a) Fig. III, 19, (b) Fig. III, 20.

2.   By using one or more line segments, convert the nontransversal subdivision S of triangle ABC shown in Fig. III, 20 into a transversal subdivision S_t of that triangle. Then prove that Theorem 46 must hold for S inasmuch as it holds for S_t.

---

* An angle of $(180/\pi)°$ at the center of a circle does not intercept an arc equal to the radius (see §14). Hence a standard unit should not be called a radian, as in Euclidean geometry.

3. Why can we not claim to have proved that the smaller the area of a triangle, the smaller its defect?

4. If a triangle has an angle-sum of 120°, find its defect in (a) degrees, (b) right angle units, (c) standard units.

5. Show how to obtain from any given triangle another which has (a) a smaller defect, (b) a larger defect.

6. (a) Given any right triangle, show how to obtain from it a triangle whose defect is twice as great. (b) What can be said intuitively about the relation of the areas of the two triangles? (c) Show that the problem proposed in (a) cannot be solved if the word "right" is omitted.

7. Does Theorem 46 hold in Euclidean geometry? Justify your answer.

10. QUADRILATERALS ASSOCIATED WITH A TRIANGLE

Each side of any given triangle *ABC* is the summit of a certain Saccheri quadrilateral having an important relation to the triangle. Consider side *BC* in Fig. III, 22, for example. Let *D*, *E* be the midpoints of the other sides,

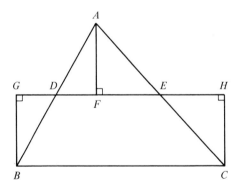

Fig. III, 22

and let *F*, *G*, *H* be the projections of *A*, *B*, *C* on line *DE*. Triangles *ADF*, *BDG* are congruent by angle–angle–side (Theo. 26), and so are triangles *AEF*, *CEH*. Hence *BG* = *AF* and *AF* = *CH*, so that *BG* = *CH*. Thus *BCHG* is a Saccheri quadrilateral with summit $\overline{BC}$. The triangle *ABC* we have used is one in which angles *ADE* and *AED* are acute, but the same result would be obtained if one of these angles is obtuse (Fig. III, 23) or if one is a right angle (Fig. III, 24).

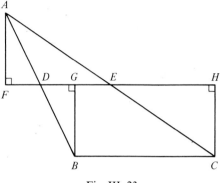

Fig. III, 23

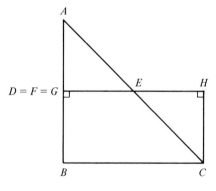

Fig. III, 24

By this same procedure a Saccheri quadrilateral with side $\overline{AB}$ as summit is obtained, and likewise one with side $\overline{AC}$ as summit. We shall speak of these three quadrilaterals as being *associated* with the triangle ABC. The following theorems are then easily proved with the aid of the congruent triangles mentioned above.

Theorem 47.   *The angle-sum of a triangle equals the sum of the summit angles in each associated Saccheri quadrilateral.*

Theorem 48.   *The line joining the midpoints of two sides of a triangle is parallel to the line containing the third side, having a common perpendicular with it, and the segment joining the midpoints is less than half of the third side.*

EXERCISES

1. Prove Theorem 47, considering all cases.

2. Prove Theorem 48 for (a) Fig. III, 22, (b) Fig. III, 23, (c) Fig. III, 24.

3. Prove anew that the angle-sum of a triangle is less than 180°.

4. Show that the perpendicular bisector of any side of a triangle is perpendicular to the line which bisects the other sides.

5. Show that a line which bisects a side of a triangle and is perpendicular to the perpendicular bisector of another side also bisects the third side.

6. Given a Saccheri quadrilateral, show how to obtain a triangle with which it is associated.

7. The Lambert quadrilateral $BCFG$ in Fig. III, 14 is said to be *associated* with the given triangle ABC; the quadrilateral and triangle have side $\overline{BC}$ in common. Similarly, there is a Lambert quadrilateral associated with this triangle so that $\overline{AC}$ is a common side. Show how to obtain this quadrilateral.

8. Given a Lambert quadrilateral, show how to obtain a triangle with which it is associated. (See Ex. 7.)

9. Verify that the sum of the acute angles in a right triangle equals the fourth angle in each associated Lambert quadrilateral. (See Ex. 7.)

10. Do Theorems 47 and 48 hold in Euclidean geometry? Justify your answer.

11. THE EQUIVALENCE OF TRIANGLES

Besides helping in the proofs of Theorems 47 and 48, the figures of the preceding section are useful in suggesting an important fact concerning area. Consider Fig. III, 22, for example. Since triangles ADF, BDG are congruent, and likewise triangles AEF, CEH, it is intuitively clear that triangle ABC has the same area as the associated Saccheri quadrilateral $BCHG$. We say "intuitively" because we have not yet defined area in hyperbolic geometry. In the absence of squares from this geometry, area can obviously no longer mean a number of square units. What it does mean is made clear in the next section. For the present let us describe in purely geometric terms the simple relation between triangle ABC and quadrilateral $BCHG$ which led us intuitively to say that they have the same area. It is that these two figures can be subdivided into parts which are congruent in pairs: triangle ADF to triangle BDG, triangle AEF to triangle CEH, and quadrilateral $BCED$ to itself. Because of this relation the two figures are said to be *equivalent*. This term can be defined in the following more general way:

Definition. Two polygons p, q are called *equivalent* if and only if they can be subdivided into the same finite number of triangles p_i, q_i, $i = 1, 2, 3, \ldots$, which are congruent in pairs, p_1 to q_1, p_2 to q_2, and so on.

In the discussion preceding this definition we did not completely subdivide triangle ABC and quadrilateral $BCHG$ into triangles as required by the definition, but this could have been done by drawing a diagonal in quadrilateral $BCED$. Hence our statement that triangle ABC and quadrilateral $BCHG$ are equivalent is correct. We obtained this result working with Fig. III, 22, but it also holds for Figs. III, 23 and III, 24, as the student can show. Thus we can state

Theorem 49.   *A triangle is equivalent to each of its associated Saccheri quadrilaterals.*

If the polygons p, q in the definition of equivalence are congruent triangles or quadrilaterals, they can obviously be subdivided into triangles which are congruent in pairs. We therefore have

Theorem 50.   *If two triangles or quadrilaterals are congruent, then they are equivalent.*

We omit the proof of the following theorem.*

Theorem 51.   *If two polygons are each equivalent to a third polygon, then they are equivalent to one another.*

We shall now state and prove two propositions which will be very helpful later in defining the area of a triangle.

Theorem 52.   *If two triangles have the same angle-sum, they are equivalent.*

Proof.   Let triangles ABC, $A'B'C'$ have the same angle-sum s. Consider, first, the case in which a side of one triangle equals a side of the other, say $BC = B'C'$. The summit angles of the Saccheri quadrilateral Q associated

---

* A proof is given in H. G. Forder, *The Foundations of Euclidean Geometry*, pp. 256–258 (Cambridge Univ. Press, London and New York, 1927); and an outline of a proof is given in H. E. Wolfe, *An Introduction to Non-Euclidean Geometry*, pp. 122, 123 (Dryden Press, New York, 1945).

with triangle ABC and having $\overline{BC}$ as summit add up to s (Theo. 47), and so do the summit angles of the Saccheri quadrilateral Q' associated with triangle $A'B'C'$ and having $\overline{B'C'}$ as summit. Their summit angles and summits being equal, Q and Q' are congruent (§8, Ex. 7), and hence equivalent (Theo. 50). They are also equivalent, respectively, to triangles ABC, $A'B'C'$ (Theo. 49). From the equivalence of triangle ABC and Q, of Q and Q', and of Q' and triangle $A'B'C'$, it follows by Theorem 51 that the two triangles are equivalent.

Now assume that no side of one triangle equals a side of the other. To be definite, suppose $A'B' > AB$. We shall consider only the case shown in Fig. III, 22, but the reasoning will hold also for the other cases. If J is a variable point on line GH (Fig. III, 25), the distance BJ is least when $J = G$,

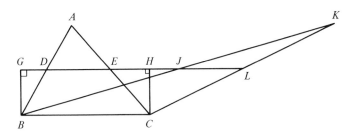

Fig. III, 25

and becomes arbitrarily great when GJ does (Theo. 18). Consider J when so located that $BJ = \frac{1}{2}A'B'$. This is possible since

$$\tfrac{1}{2}A'B' > \tfrac{1}{2}AB = BD > BG.$$

Take point K ($\neq B$) on line BJ so that $JK = BJ$. Since C, K are on opposite sides of line GH, side CK of triangle BCK meets line GH in a point L. The perpendicular bisector of $\overline{BC}$ in Saccheri quadrilateral $BCHG$ is perpendicular to $\overline{GH}$ (Theo. 30). Since line GH contains the midpoint J of side BK in triangle BCK and is perpendicular to the perpendicular bisector of side BC, it also contains the midpoint of side CK (§10, Ex. 5). Hence L is the midpoint of side CK.

Thus Saccheri quadrilateral $BCHG$ is associated with triangle BCK, as well as with triangle ABC, and hence is equivalent to each of these triangles (Theo. 49). By Theorem 51, then,

$$\text{triangles } ABC, BCK \text{ are equivalent.} \tag{1}$$

Also, being associated with the same quadrilateral,

$$\text{triangles } ABC, BCK \text{ have the same angle-sum} \tag{2}$$

by Theorem 47. From (2) and our hypothesis that triangles ABC, $A'B'C'$ have the same angle-sum, we infer that

$$\text{triangles } A'B'C', \, BCK \text{ have the same angle-sum.} \qquad (3)$$

Since $BK = A'B'$, it follows from (3) and the first part of the proof that triangles $A'B'C'$, BCK are equivalent. On combining this with (1) and using Theorem 51, we conclude that triangles ABC, $A'B'C'$ are equivalent.

The converse of Theorem 52 is also true, as we now show.

Theorem 53.   *If two triangles are equivalent, they have the same angle-sum.*

Proof.   Let triangles ABC, $A'B'C'$ be equivalent. By definition, then, they can be subdivided into the same finite number of triangles that are congruent in pairs. The two triangles in each pair, being congruent, have the same angle-sum, and hence the same defect. Triangles ABC, $A'B'C'$ then have the same defect (Theo. 46), and therefore the same angle-sum.

Combining Theorems 52 and 53, we see that two triangles are equivalent if and only if they have the same angle-sum.

EXERCISES

1.   Prove Theorem 49 for the cases shown in (a) Fig. III, 23, (b) Fig. III, 24.

2.   If polygons p, q are equivalent to polygons r, s, respectively, and the latter are mutually equivalent, then so are p, q. Prove this.

3.   If two triangles are equivalent, they are not necessarily congruent. This was seen in the proof of Theorem 52 (triangles ABC, BCK in Fig. III, 25), but can also be shown in a simpler situation. Let ABC be an isosceles triangle with a right angle at C. By means of a suitable transversal from C, triangle ABC can be subdivided into two triangles which, upon rearrangement, will form a new triangle equivalent to the original but not congruent to it. Show this.

4.   Prove that two Saccheri quadrilaterals are equivalent if a summit angle of one equals a summit angle of the other.

5.   Show that a right triangle is equivalent to each associated Lambert quadrilateral. (See §10, Ex. 7.)

6.   Prove that two Lambert quadrilaterals are equivalent if their acute angles are equal. (Use the preceding exercise and §10, Exs. 8, 9.)

7. The definition of equivalent polygons and Theorems 49 through 51 hold in Euclidean geometry, but Theorem 52 does not. Find the step in the proof of Theorem 52 that accounts for this.

12. AREA OF A TRIANGLE

Our familiarity with areas in daily life leads us to expect that the area of a triangle in hyperbolic geometry, as in Euclidean geometry, should be a number which measures the inner content of the triangle, that is, the part of the plane enclosed by the triangle. As we have already noted, this number cannot be a number of square units since squares do not exist in hyperbolic geometry. What can this number be then? It might conceivably be a number of *equilateral triangle units* since equilateral triangles do exist in hyperbolic geometry, or a number of *regular hexagon units*, and so forth. Or it might very well be a number suggested by the concept of equivalence, which is clearly concerned with the inner content of geometrical figures. Let us look into this latter approach first since it is the easier and the more promising.

In the preceding section we saw how natural it is to say, speaking intuitively, that a triangle and each associated quadrilateral have the same area since they are equivalent. It is also natural to expect that any two equivalent polygons have the same area. Therefore we should try to define the area of a triangle so that two equivalent triangles always have the same area. Now, according to Theorems 52 and 53, two triangles are equivalent if and only if they have the same angle-sum. We should therefore try to define area so that two triangles have equal areas if and only if they have equal angle-sums. A simple way of doing this is to take the areas of triangles to be numbers proportional to their angle-sum numbers, or, in more familiar language, to define the area of a triangle to be proportional to its angle-sum. Then, for example, triangles with the angle-sums 50°, 100°, 150° would have the areas $50k$, $100k$, $150k$, where k is any positive constant.

It is easy to show, however, that this definition violates one of our expectations concerning area. For consider a triangle ABC (Fig. III, 26) which

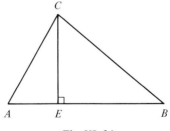

Fig. III, 26

is subdivided into two right triangles by the perpendicular $\overline{CE}$. It would be reasonable to expect that

$$\text{area } ACE + \text{area } BCE = \text{area } ABC, \tag{1}$$

but this is impossible under the proposed definition. For, triangles ACE and BCE, having angle-sums exceeding 90°, would have areas exceeding 90k according to the definition, and hence a combined area exceeding 180k, whereas triangle ABC, with an angle-sum less than 180°, would have an area less than 180k.

There are many other simple definitions we could try, for the condition that two triangles with equal angle-sums should have equal areas merely requires the area to be a function of the angle-sum, that is, to each angle-sum number there should correspond a unique area number. We could therefore try defining the area to be proportional to the square of the angle-sum, or to its cube, or to its square root, and so forth, and trust that we would hit upon a functional relation which satisfies equation (1).

Fortunately, there is a more satisfactory solution of the problem. It will be recalled that what led us to the unsuitable definition was the basic fact that two triangles are equivalent if and only if they have the same angle-sum. Now, this fact can also be expressed by saying that two triangles are equivalent if and only if they have the same defect. When this is done, we expect triangles to have the same area only if they have the same defect. As before, a simple way to achieve this is to define area to be proportional to defect, meaning, of course, that the area numbers are proportional to the defect numbers. The difficulty encountered with our first definition does not occur now. For suppose the defects of the triangles in equation (1), from left to right, are a, b, c. Then, under the new definition, the areas of these triangles would be ka, kb, kc. Since the defect of triangle ABC equals the sum of the defects of triangles ACE and BCE (Theo. 46), we have

$$a + b = c.$$

Multiplying by k gives

$$ka + kb = kc,$$

or

$$\text{area } ACE + \text{area } BCE = \text{area } ABC.$$

Thus the new definition satisfies equation (1). Clearly, it will satisfy (1) even when $\overline{CE}$ is not perpendicular to $\overline{AB}$.

The property expressed in equation (1), where E is any point between A and B, is only a special case of a more general property of area demanded by intuition, namely, that the area of a triangle should always equal the sum of the areas of the triangles resulting from a subdivision of the triangle. Clearly, in view of Theorem 46, this demand is met by the new definition. The definition therefore meets all our expectations concerning area and we can now adopt it formally.

Definition.   The *areas of triangles* are numbers proportional to the defect numbers of the triangles, the constant of proportionality being the same for all triangles. In more familiar language, the area A of a triangle is proportional to its defect D,

$$A = kD, \qquad\qquad (2)$$

where k is any positive constant.

For example, when degree measure is used, triangles with angle-sums of $50°$, $100°$, $150°$, and hence with defects of 130, 80, 30, have the areas 260, 160, 60 if $k = 2$, and the areas 26, 16, 6 if $k = \frac{1}{5}$. This somewhat resembles the Euclidean situation, where the area of a triangle depends on the choice of linear unit, that is, the unit of length. In fact, it can be shown that different values of k correspond to different linear units. Thus the area of a triangle in hyperbolic geometry depends on the choice of both the linear unit and the angular unit. In Euclidean geometry the area depends only on the linear unit.

One question remains unanswered. Knowing that any formula expressing A as a function of D, where D and A are positive, would guarantee that triangles have equal areas only if their defects are equal, we tried the simple formula $A = kD$, found that it satisfies equation (1), and therefore adopted it as our definition. What if some other formula also satisfies (1)? This cannot happen. We leave it to Exercise 6 to show why.

According to formula (2), two triangles will have the same area if and only if they have the same defect, and hence if and only if they have the same angle-sum. Thus we have

Theorem 54.   *Two triangles have the same area if and only if they have the same angle-sum.*

EXERCISES

1.   Using degree measure and $k = 2$, find the area of a triangle whose angle-sum is $160°$.

2.   Using $k = 2$, find the area of the triangle in Exercise 1 corresponding to each of the following angular units: (a) a complete rotation, (b) a right angle, (c) a standard unit.

3.   Prove that the area of a triangle t is equal to the sum of the areas of the triangles resulting from any subdivision of t into triangles.

4. If the defects of the triangles in equation (1), from left to right, are a, b, c, show that (1) is not satisfied by the simple formula $A = D + 1$, where A, D mean the same as in equation (2).

5. Do Exercise 4 using the simple formula $A = 2D + 1$.

6. Verify each step of the following argument, which shows that $A = kD$ is the only relation between D and A satisfying equation (1). Assume that $A = f(D)$ is an unknown relation between A and D that satisfies (1). Let the defects of the triangles in (1) from left to right be a, b, c. Then $a + b = c$ (Theo. 46). Since $A = f(D)$ satisfies (1), we also have $f(a) + f(b) = f(c)$. Holding in all cases, these equations must hold when $a = b$. It follows that $f(2a) = 2f(a)$. This equation shows that the area is doubled when the defect is doubled. Similarly, $f(3a) = 3f(a)$, $f(4a) = 4f(a)$, and so forth. As intuition then suggests, the area is proportional to the defect. This can be proved.* Thus the unknown relation $A = f(D)$ is necessarily $A = kD$.

7. We gave up trying to define the area of a triangle in terms of its angle-sum because it seemed that the correct formula could only be discovered by trial and error. Find the formula now by deducing it from equation (2).

13. IMPLICATIONS OF THE AREA FORMULA

Having defined area numbers of triangles, let us return to a question raised at the beginning of the preceding section. Since area numbers are not numbers of square units, could they perhaps be numbers of some other kind of unit? To look into this as simply as possible, let us take $k = 1$ in our area formula and use degree measure. Triangles with defects 1°, 2°, 3°, ... then have areas 1, 2, 3, ..., and vice versa. Consider a triangle ABC (Fig. III, 27) with area 2, defect 2°, and angle-sum 178°. Let D be a point on side AB that varies from A to B. Then angles DCB, CDB will vary and, if we suppose this variation is continuous, $\measuredangle DCB$ will approach 0° when D approaches B, and $\measuredangle CDB$ will approach the supplement of $\measuredangle ABC$. Accordingly, the angle-

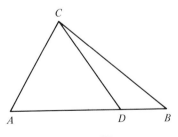

Fig. III, 27

* The student is not expected to give the proof.

sum of triangle BCD will vary, starting with the value 178° and approaching 180° when D approaches B. Consequently there will be some position of D for which this angle-sum is 179°. The defect of triangle BCD will then be 1°, so that the defect of triangle ADC must also be 1° (Theo. 46). Triangle ABC, with area 2, is thus divided into two triangles each with area 1. Regarding the latter triangles as units of area, we may then say that triangle ABC contains 2 units of area.

Similarly, a triangle with area 3 can be subdivided into three triangles each of area 1, and hence may be said to contain 3 units of area. The argument clearly applies to any triangle whose area is an integer. The extension of the argument to a triangle whose area is not an integer presents no problem. For example, a triangle with area $3\frac{1}{2}$ is subdivisible into four triangles, of which three have area 1 and one has area $\frac{1}{2}$, and hence can be said to contain $3\frac{1}{2}$ units of area. Since triangles of area 1 thus serve as units of area, we may call them *triangular units* or *unit triangles*. Unlike the unit squares of Euclidean geometry, these unit triangles are, in general, not congruent.

We have been considering only the case $k = 1$, but the result is the same for any value of k; that is, the area number of a triangle can always be regarded as a number of triangular units of area, a triangular unit being a triangle of area 1. Clearly, the defects of the unit triangles are 1° only when $k = 1$.

We now consider another important implication of the area formula, one concerning the extreme values which areas and angle-sums of triangles can have. Let us use a specific value of k in the area formula, say 5, so that

$$A = 5D.$$

Since angle-sums of triangles lie between 0° and 180°, the corresponding defects are between 180° and 0°, and the values of D are between 180 and 0. Hence A must lie between 900 and 0. In other words, with our choice of angular unit and k, no triangle can have an area as great as 900, no triangle can contain as much as 900 units of area. This fact is striking when we realize that there is no limit to the lengths of the sides of a triangle. A right triangle, for example, might have arms each a million units long, and hence a hypotenuse even longer, and yet its area would be less than 900. Clearly, it is a characteristic of the formula $A = kD$ that each choice of k and angular unit leads to a certain upper bound for the areas of triangles. This contrasts with the Euclidean situation, where the formula *area* $= \frac{1}{2}$ *base* $\times$ *altitude* permits areas to exceed all bounds regardless of the choice of units.

In the example above, the closer A is to 900, the closer D will be to 180, and hence the closer will the angle-sum be to 0°. On the other hand, the closer A is to 0, the closer will D be to 0, and hence the closer will the angle-sum be to 180°. Thus triangles with very large areas have angle-sums greatly different from what they are in Euclidean geometry, but triangles with very small areas have angle-sums only slightly less than what they are in Euclidean geometry. We have here our first illustration of the general fact that (1) the smaller we take a figure in hyperbolic geometry the more closely, in the main, do its metric properties (those involving measurement) agree with those of the same-named figure in Euclidean geometry, whereas (2) by taking a figure sufficiently large we can obtain radical departures in its metric properties from those in the corresponding Euclidean situation. The content of (1) is often expressed briefly by saying that, *in the small*, or *locally*, hyperbolic geometry is approximately Euclidean.

To offer another illustration of this general fact, let us consider a Saccheri quadrilateral of comparatively small area, say one associated with a triangle whose angle-sum is 179°59′. (Strictly we have not defined the area of this quadrilateral, but it is intuitively clear that the quadrilateral, being equivalent to the triangle, has the same small area.) By Theorem 47 the summit angles of the quadrilateral are each 89°59′30″, which is very close to the 90° value which a summit angle always has in Euclidean geometry. Clearly, we would come even closer in Saccheri quadrilaterals of smaller area. Conversely, a Saccheri quadrilateral with summit angles very close to 90°, being associated with a triangle whose angle-sum is very close to 180°, would have a very small area. Looking at things the other way, a Saccheri quadrilateral of comparatively large area would have summit angles differing greatly from 90°. Thus, a quadrilateral which is associated with a triangle of angle-sum 2°, and hence which is very large, has summit angles each of 1°.

Since it is a familiar fact that the conventional drawings made on paper or blackboard with compass, straightedge, protractor, and so forth, are accurate representations of situations in Euclidean geometry, the preceding discussion shows that such drawings cannot also accurately represent situations in hyperbolic geometry unless the figures dealt with are small enough so that their metric properties are approximately Euclidean. A figure whose metric properties differ greatly from Euclidean ones, and hence which is very large, can be represented on a sheet of paper only by being distorted. This was already noted in Section 7. Thus, if we are exhibiting a triangle with an angle-sum of 150° and want the angles in the drawing to add up to this value, we must make the sides look slightly curved, as in Fig. III, 28.

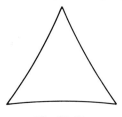

Fig. III, 28

EXERCISES

1. Using degree measure and $k = 1$, show that a triangle with area 3 can be divided into three triangles each with area 1.

2. Using degree measure and $k = \frac{1}{3}$, show that the unit triangles have defects of $3°$ and that a triangle with area 2 can be divided into two triangles each with area 1.

3. What is the relation between the value of k and the defect number of a unit triangle?

4. (a) Given any right triangle ABC, show how to form a triangle whose area is twice as great and prove you are correct. (b) Since triangle ABC can always be taken larger and larger, say by extending the arms further, why can we not conclude from (a) that triangular areas can be arbitrarily great?

5. In view of Theorem 49, we define the area of a Saccheri quadrilateral to equal the area of any triangle with which it is associated. Show that two Saccheri quadrilaterals have the same area if and only if a summit angle of one equals a summit angle of the other.

6. In view of Section 11, Exercise 5, we define the area of a Lambert quadrilateral to equal the area of any right triangle with which it is associated. Show that two Lambert quadrilaterals have the same area if and only if their acute angles are equal. (See §10, Ex. 9.)

7. Noting that the formula for the area of a triangle with angles $\lambda°$, $\mu°$, $\nu°$ can be written

$$A = k(180 - \lambda - \mu - \nu),$$

show that the formula for the area of a Saccheri quadrilateral with a summit angle of $S°$ is $A = 2k(90 - S)$. (See Ex. 5.)

8. Noting that the formula for the area of a right triangle with acute angles $\lambda°$, $\mu°$ can be written $A = k(90 - \lambda - \mu)$, show that the formula for the area of a Lambert quadrilateral with a fourth angle of $F°$ is $A = k(90 - F)$. (See Ex. 6.)

9. When suitably defined, the area of a quadrilateral with angles $\alpha°$, $\beta°$, $\gamma°$, $\delta°$ is given by the formula $A = k(360 - \alpha - \beta - \gamma - \delta)$, where k is the same constant as in the formula for the area of a triangle. Give a suitable definition. (See Ex. 7.)

10. Show that the formula for the area of the quadrilateral given in Exercise 9 is consistent with the formulas for the areas of Saccheri and Lambert quadrilaterals given in Exercises 7 and 8.

11. When the defect of a quadrilateral is suitably defined, the area of the quadrilateral as given by the formula in Exercise 9 is proportional to the defect. Give a suitable definition.

12. Using a straightedge and compass, carefully construct a Saccheri quadrilateral and measure the base, summit, and summit angles. Are your results consistent with the facts of hyperbolic geometry? If they are not, does this necessarily mean that hyperbolic geometry disagrees with experience? Explain.

13. Make drawings to illustrate the following: (a) a Saccheri quadrilateral with summit angles of 60°; (b) a Lambert quadrilateral whose fourth angle is 30°; (c) a triangle each angle of which is 45°.

14. In hyperbolic geometry, as in Euclidean geometry, straight lines are "shortest paths," that is, the shortest path joining two points is the straight line segment connecting the points. Prove this to the extent of showing that the straight line segment is shorter than any other rectilinear path joining the points, that is, any other path consisting of straight line segments.

14. CIRCLES

As in Euclidean geometry, a *circle* is defined to be the set of all points which are at a constant distance from a fixed point, the distance and fixed point being called the *radius* and the *center* of the circle. Likewise all the other familiar terms connected with circles (diameter, arc, chord, secant, tangent, central angle, inscribed angle, congruent circles, etc.) retain their usual meaning. It follows that all the familiar theorems on circles whose proofs make no direct or indirect use of Euclid's Postulate 5 still hold in hyperbolic geometry. Thus, for example, the line joining the center of a circle to the midpoint of a chord is perpendicular to the chord, a central angle is measured by (that is, is proportional to) its intercepted arc, and in the same circle or congruent circles equal central angles subtend equal chords.

It is no longer true, however, that an angle inscribed in a semicircle is a right angle, that an inscribed angle is measured by half its intercepted arc, or that through any three noncollinear points there passes a circle, for these properties are consequences of Postulate 5. Further, while it is still true that there are 360° in a circle and that all central angles of 1° in a given circle subtend equal arcs, no longer is the circumference 2π times the radius. Hence a central angle of 1 standard unit (or $180/\pi$ degrees) does not subtend an arc equal to the radius, as it does in Euclidean geometry. We shall therefore not refer to a standard unit as a radian.

Before proceeding to the proof of two new theorems on circles, let us recall that a tangent at a point P of a circle with center O is defined to be the line through P perpendicular to the segment (or radius) $\overline{OP}$. It can be proved that the tangent contains no point of the circle other than P (Ex. 1).

Theorem 55. *The angles inscribed in a semicircle of any given circle are acute and do not have a constant value.*

Proof. Consider a circle with center O and diameter $\overline{AB}$ (Fig. III, 29).

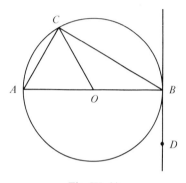

Fig. III, 29

If C is any point of the circle other than A or B, then $\angle ACB$ is inscribed in a semicircle. From triangle ABC we have

$$\angle A + \angle B + \angle ACB < 180°. \tag{1}$$

Since triangles OAC, OBC are isosceles, $\angle A = \angle OCA$ and $\angle B = \angle OCB$, so that

$$\angle A + \angle B = \angle OCA + \angle OCB = \angle ACB.$$

Substituting this in (1) gives

$$2(\angle ACB) < 180°$$
$$\angle ACB < 90°.$$

To prove the last part of the theorem, let line BD be tangent to the circle at B. If A, B are kept fixed and C varies, approaching B, lines AC and BC will approach lines AB and BD. Hence $\angle ACB$ will approach $\angle ABD$, which is a right angle. It follows from this that if $\angle ACB$ had a constant value for all positions of C (other than A and B), the value would have to be 90°. This would contradict the first part of the theorem. Hence $\angle ACB$ does not have a constant value.

Since the measure of a semicircle is 180° (that is, a semicircle subtends a central angle of 180°), we see from Theorem 55 that the Euclidean proposition, *an inscribed angle is measured by half its intercepted arc*, does not hold in hyperbolic geometry, as was mentioned earlier.

The Euclidean proposition, *through any three noncollinear points there passes a circle*, is logically equivalent to Postulate 5 and is listed among the substitutes for the postulate in Chapter I, Section 6 [see statement (i)]. We can therefore conclude immediately that the proposition does not hold in hyperbolic geometry. However, it will be worthwhile to give a detailed proof of this since we never demonstrated that statement (i) is a substitute.

Theorem 56. *There does not always exist a circle passing through three given noncollinear points.*

Proof. Let us first note that if three points lie on a circle, then they are necessarily the vertices of a triangle, and the perpendicular bisectors of the sides of the triangle pass through the center of the circle. (This is easily shown, just as in Euclidean geometry, and is left as Ex. 2.) Hence to prove the present theorem we need only exhibit three points which are the vertices of a triangle two of whose sides are bisected at right angles by a pair of parallel lines, for no circle can pass through three such points. To do this we take two parallel lines a, b which have a common perpendicular c (Fig. III, 30). Choose any point A not on a, b, or c. Let P and Q be the projections of A on a and b. Extend $\overline{AP}$ to B so that $PB = PA$, and $\overline{AQ}$ to C so that $QC = QA$. The points A, B, C do not lie on a line, for if they did, the line would be perpendicular

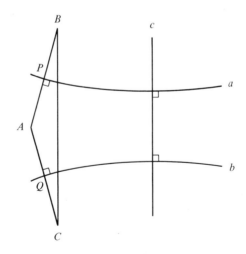

Fig. III, 30

to *a* and *b*, yet different from *c*, which is impossible. Thus, *A*, *B*, *C* are the vertices of a triangle two of whose sides, $\overline{AB}$ and $\overline{AC}$, are bisected at right angles by the parallel lines *a*, *b*. As noted earlier in the proof, no circle can pass through the points *A*, *B*, *C*.

We are in no position to state the formulas for the circumference and area of a circle (they are considered in VI, §8). As in Euclidean geometry, limits are required to obtain these formulas, the circumference being the limit approached by the perimeter of an inscribed regular polygon of *n* sides when *n* becomes infinite, and the area being the limit approached by the area of that polygon. The formulas are not as simple as in Euclidean geometry, the ratio of the circumference to the diameter, for example, no longer being constant.

EXERCISES

1. Prove that the tangent to a circle meets it in no point other than the point of tangency.

2. If *A*, *B*, *C* are three points on a circle with center *O*, prove that the perpendicular bisectors of the sides of triangle *ABC* go through *O*.

3. Show that the bisectors of the angles of a triangle are concurrent (that is, meet in a point), and hence that a circle can always be inscribed in a given triangle.

4. Prove that an angle inscribed in a circle is less than half the central angle subtending the same arc, or, as is sometimes said, an inscribed angle is measured by less than half its intercepted arc. In doing this, consider these positions of the center of the circle: (a) on a side of the angle, (b) inside the angle, (c) outside the angle. (Use §8, Ex. 11.)

5. Show that if a quadrilateral is inscribed in a circle, the sum of one pair of opposite angles equals the sum of the other pair. Is this also true in Euclidean geometry?

6. Show that if the perpendicular bisectors of two sides of a triangle meet, then the perpendicular bisector of the third side goes through their meeting point and the vertices lie on a circle.

7. If the perpendicular bisectors of two sides of a triangle are parallel, with a common perpendicular, then the perpendicular bisectors of all three sides are parallel, with the same common perpendicular. Prove this, using Fig. III, 31, where *ABC* is the given triangle, lines *A'H*, *B'I* are the given perpendicular bisectors, line *HI* is their common perpendicular, and *C'* is

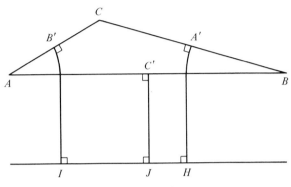

Fig. III, 31

the midpoint of $\overline{AB}$. (Let J be the projection of C' on line HI and show that line $C'J$ is also perpendicular to $\overline{AB}$.)

8.   Show that triangles do exist in which the perpendicular bisectors of the three sides are concurrent.

9.   Show that an equidistant curve has the property that no three of its points are on a circle. (See §5, Exs. 6, 7.)

10.   Given a circle with center O and radius r, and a point P outside of the circle, show how to construct a line through P which is tangent to the circle. Is the construction also valid in Euclidean geometry?

11.   Show that the circumference of a circle in hyperbolic geometry is more than six times the radius.

12.   Why is it inappropriate in hyperbolic geometry to call an angle of $180/\pi$ degrees a radian?

IV

Parallels Without
a Common Perpendicular

1. INTRODUCTION

Thus far we have worked with just one type of parallel lines, those possessing a common perpendicular, and have developed only such parts of hyperbolic geometry as involve them. There remain, however, many important aspects of the subject which depend on the other type of parallels, those not possessing a common perpendicular, and we shall now consider them.

2. PARALLELS WITHOUT A COMMON PERPENDICULAR

It will be recalled that one of the things correctly deduced by Saccheri from his hypothesis of the acute angle was the existence of parallels without a common perpendicular (II, §5). Since we are using his hypothesis of the acute angle as our hyperbolic parallel postulate, it follows that parallels without a common perpendicular exist in hyperbolic geometry. Although there is nothing wrong with this line of reasoning, the student may be skeptical of it inasmuch as we really never examined Saccheri's proof. We shall therefore now show, independently of Saccheri's work, that parallels without a common perpendicular exist. At the same time we shall see how many such parallels there are and what relation they have to parallels with a common perpendicular.

If g is any line and F any point not on it (Fig. IV, 1), we know from Chapter III that there are infinitely many lines through F which are parallel to g and have a common perpendicular with it (III, §7, Theo. 40). More precisely, if E

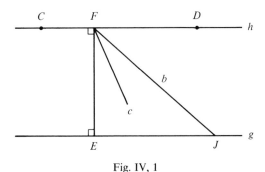

Fig. IV, 1

is the projection of F on g, and h is the line through F at right angles to $\overline{EF}$, there are infinitely many lines subdividing $\measuredangle EFD$ which have a common perpendicular with g, infinitely many subdividing $\measuredangle EFC$ with this same property, and in each of these sets of lines there is no line that is closest to line EF in the sense of making a smaller angle with it than do all the others (III, §7, Theo. 41). We shall now show that in addition to these parallels to g through F there are two others through F which do not have a common perpendicular with g.

Consider the set $\{l\}$ of *all* the lines which subdivide $\measuredangle EFD$. If we take any two of them, then one of the two will make a smaller acute angle with line EF than does the other. Then we can say that this one *precedes* the other (III, §2, Property 6). Now, the subdividers $\{l\}$ fall into two classes S and T: in S are the subdividers which meet g, and in T are the subdividers which do not meet g. Also, each member of S precedes each member of T. (For suppose this is not true and that some line c in T precedes some line b in S, as shown in Fig. IV, 1. Then c makes a smaller angle with line EF than does b. Since b meets g in a point J, line c subdivides $\measuredangle EFJ$ of triangle EFJ and hence meets side EJ. This contradicts that c does not meet g.)

It now follows that there is a subdivider (call it m) which is the boundary between the members of S and those of T (III, §2, Property 14). This means that if m is in S, then it is preceded by every other member of S; and if it is in T, then it precedes every other member of T. Actually, m must be in T, for assume it is in S. Then it meets g in some point K (Fig. IV, 2). Take any point L on g such that K is between L and E. Then line FL is in S and is preceded by m since m makes a smaller angle with line EF than does line FL. But, being in S and also being the boundary line, m must be preceded by every other line in S, as mentioned above. From this contradiction we conclude that m is in T.

Thus m is a line which does not meet g and precedes all the other lines with this property. It therefore makes a smaller angle with line EF than do all these others. By Theorem 41 in Chapter III, then, it cannot have a common

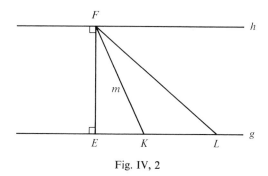

Fig. IV, 2

perpendicular with g. An analogous discussion dealing with the lines sub-dividing $\angle EFC$ (Fig. IV, 3) would show that one of these lines, n, is a boundary line, does not meet g, and has no common perpendicular with g. At the end of Section 7 we shall prove that m and n are the only lines through F which do not meet g and have no common perpendicular with it.

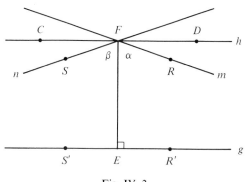

Fig. IV, 3

The acute angles α, β (Fig. IV, 3) which m, n make with line EF are equal. For assume $\alpha > \beta$. Let line AF subdivide α so that $\angle EFA = \beta$ (Fig. IV, 4). Then line AF meets g in a point J. Take J' on g so that E' is midway between J and J'. From the congruence of triangles EFJ, EFJ' we conclude that $\angle EFJ' = \angle EFJ = \beta$, and hence that line FJ' coincides with n. Since this is impossible, α cannot exceed β. Similarly, β cannot exceed α. Thus $\alpha = \beta$.

We summarize our results:

Theorem 1. *Given any line g and any point F not on it, there exist exactly two lines m, n which go through F, are parallel to g, and do not have a common perpendicular with g. If E is the projection of F on g, then m, n make equal acute angles α, β with line EF. Within one pair of vertical angles formed by m and n*

(namely, the pair containing line EF) lie all the lines through F which meet g; within the other pair of vertical angles lie all the parallels to g through F which have a common perpendicular with g.

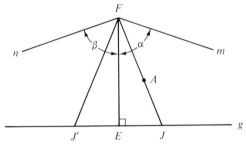

Fig. IV, 4

In recognition of their boundary character we shall call *m* and *n* the *boundary parallels* to *g* through *F*. The other parallels to *g* through *F* will be called *nonboundary parallels*. The equal angles α, β are called the *angles of parallelism* corresponding to *F* and *g*. The sides of these angles which lie on *m* and *n* will be called *boundary rays*. Thus, ray *FR* is the boundary ray on *m* and ray *FS* is the boundary ray on *n* (Fig. IV, 3).

As we have seen, line *EF* and each line subdividing an angle of parallelism will meet *g*. This property can be extended in a simple way that is often useful. If we take any point *P* on *g* (Fig. IV, 5) and any point *Q* on a boundary ray, it is clear that any line which subdivides ∡*PFQ* will meet *g*. We state this formally:

Theorem 2.   *Consider the boundary parallels to g through F. If P is any point of g, and Q is any point on a boundary ray, then every line subdividing* ∡*PFQ meets g.*

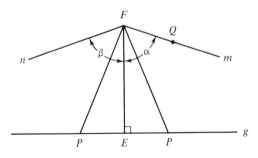

Fig. IV, 5

It is clear that $\angle PFQ$ may be acute, obtuse, or a right angle depending on the position of P, and that it is not an angle of parallelism unless P coincides with E (Fig. IV, 5). The following converse of Theorem 2 is also true and very helpful in identifying boundary parallels. We leave its proof as Exercise 3.

Theorem 3.   *If two nonintersecting lines m, g are met by a transversal in points F, P, and every line subdividing one of the interior angles* at F meets g, then m is a boundary parallel to g through F.*

The fact that angles of parallelism are acute enables us to prove the following:

Theorem 4.   *If one line is a boundary parallel to another through a point, there is a transversal which meets them so that the interior angles on the same side of the transversal are equal.*

Proof.   Let m be a boundary parallel to g through F (Fig. IV, 6), E the projection of F on g, and α the corresponding angle of parallelism in degrees. Let D be the projection of E on m, and γ the degree measure of the acute angle CED, where C is on g as shown in Fig. IV, 6. Let P be a variable point which

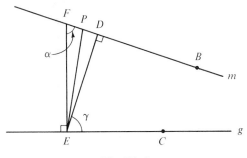

Fig. IV, 6

may take all positions on $\overline{DF}$. Consider the value of $\angle EPB - \angle PEC$. It varies continuously when P varies from D to F, is equal to the positive number $90 - \gamma$ when $P = D$, and is equal to the negative number $\alpha - 90$ when $P = F$. It must therefore be zero for some position of P between D and F. For that position we have $\angle EPB = \angle PEC$.

It is often useful to be able to distinguish between the two boundary parallels to g through F. Consider Fig. IV, 3, for example, where g is hori-

* The ray FP is one side of each interior angle at F and the other side is on m.

zontal and line EF is vertical. Noting that the boundary ray on m (and also the interior of α) lies to the right of line EF, whereas the boundary ray on n (and also the interior of β) lies to the left of line EF, we may call m and n the *right-hand* and *left-hand* parallels to g through F. These names can also be justified in another way. Let us first agree that the *direction of any ray PQ* shall mean the direction from the endpoint P toward Q. Now consider the boundary rays FR, FS on m, n (Fig. IV, 3). Their projections on g are the rays ER', ES'. The direction of ray ER' is to the right on g, and the direction of ray ES' is to the left on g. Thus we may call m and n the right-hand and left-hand parallels because their boundary rays project into rays on g which extend to the right and left, respectively.

The boundary parallels to g through any point other than F can, of course, be designated in the same way. We may therefore speak of all right-hand parallels to g as being parallel to g in the *same direction* (meaning on g), do likewise for all left-hand parallels, and also regard any left-hand and any right-hand parallel to g as being parallel to g in *opposite directions* (meaning on g). In Fig. IV, 7, for example, lines m, p, r are right-hand parallels to g,

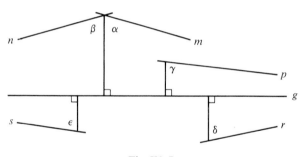

Fig. IV, 7

the corresponding angles of parallelism being α, γ, δ, and lines n, s are left-hand parallels to g, with angles of parallelism β, ε. Lines m, p, r are parallel to g in one direction (on g) and lines n, s are parallel to g in the other, or opposite, direction (on g).

When g is not horizontal the terms "right" and "left" may, of course, be inappropriate, but we can still speak of boundary parallels to g as being in the same or opposite directions. In Fig. IV, 8, for example, n and p are parallel to g in the same direction (on g), with angles of parallelism β and γ, whereas m and n are parallel to g in opposite directions (on g), as are m and p. The justification for this terminology is, as before, simply that any two boundary rays will project into rays on g which extend either in the same direction on g or in opposite directions.

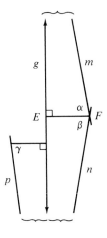

Fig. IV, 8

The directions on g which we have been discussing are known as *directions of parallelism*. Braces and arrows will be used in our diagrams to identify boundary parallels and indicate the directions of parallelism. In Fig. IV, 8, for example, m and g are bracketed and there is an upward-pointing arrow on g near the brace. This means that m is parallel to g in the upward direction on g. Similarly, we interpret the diagram to mean that p is parallel to g in the downward direction on g.

For brevity we have often omitted the word "boundary," saying "right-hand parallel" instead of "right-hand boundary parallel," "parallel in the same direction" rather than "boundary parallel in the same direction," and so forth, and shall continue doing this.

EXERCISES

1. If g, h are two lines which do not meet, state whenever possible in each of the following cases whether h is a boundary parallel to g or a non-boundary parallel:

(a)   no perpendicular to h meets g at right angles;

(b)   there is a Saccheri quadrilateral whose base and summit are on g and h;

(c) g, h have a transversal forming equal alternate interior angles;

(d)   the distances to g from two points of h are always unequal;

(e) g, h have a transversal forming a pair of alternate interior angles of which one is acute and the other right;

(f) g, h have a transversal such that all of the lines subdividing one of the interior angles at h meet g.

2. Illustrate the following in one diagram: h goes through F parallel to the vertical line g, with angle of parallelism α; h and j are parallel to g in opposite directions; k is a nonboundary parallel to g through F.

3. Prove Theorem 3.

3. PROPERTIES OF BOUNDARY PARALLELS

Several basic questions concerning boundary parallels arise immediately. If h is a boundary parallel to g, does it follow that g is a boundary parallel to h? Of course, g is parallel to h since it does not meet h, but is it the special kind of parallel which we have called boundary parallel? Another question is this: If h and j are boundary parallels to g in the same direction, are they boundary parallels to each other? A third question is the following: If h is a boundary parallel to g through P, and P' is any other point on h, will h also be a boundary parallel to g through P'? As in the first question, it is clear that h will be a parallel to g through P', but not clear without proof that it is a boundary parallel. We now give the proof.

Theorem 5. *If a line is the parallel, through one of its points, to another line in a given direction, then it is the parallel, through each of its points, to that line in that same direction.*

Proof. Let g be any line and P any point not on it. For convenience we take g to be horizontal. Let h be the right-hand parallel to g through P (Fig. IV, 9). Let P' be any other point of h, and Q, Q' the projections of P, P' on g. We shall show that h is the right-hand parallel to g through P'. Suppose, first, that P' is to the right of P (Fig. IV, 9). Since we already know

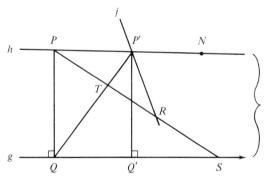

Fig. IV, 9

that h does not meet g, it remains to show that every line j subdividing $\sphericalangle NP'Q$ meets g, where N is on h to the right of P'. Let R be a point of j between h and g. Being interior to $\sphericalangle NP'Q'$, R is necessarily interior to $\sphericalangle NPQ$. Hence line PR subdivides $\sphericalangle NPQ$. It therefore meets g in some point S since h is the right-hand parallel to g through P. Line PR, subdividing the angle at P in triangle PQP', must meet side $P'Q$ of the triangle in a point T between P' and Q (III, §2, Property 11). Considering triangle QTS, we note that j meets side ST in R, cannot meet side QT, and must therefore meet side QS (Axiom of Pasch). Thus j meets g.

Suppose, now, that P' is to the left of P (Fig. IV, 10). We must show that any line j subdividing $\sphericalangle PP'Q$ meets g. Take R on j above h. Then R is interior to $\sphericalangle MP'T'$, and hence interior to $\sphericalangle P'PT$. Therefore line PR subdivides $\sphericalangle P'PT$, and hence also $\sphericalangle NPQ$. It then follows, since h is the

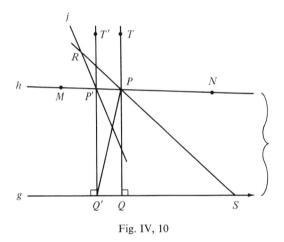

Fig. IV, 10

right-hand parallel to g through P, that line PR meets g in some point S. Considering triangle $PP'Q'$, we see that j must meet side PQ'. Then, considering triangle $PQ'S$, we note that j cannot meet side PS (since it meets line PS in R), and hence must meet side $Q'S$ (Pasch's Axiom). Thus j meets g.

In view of the preceding theorem we may speak of one line h being parallel to another line g in a given direction without mentioning any particular point on h. This is because every point of h is the endpoint of a boundary ray. It should be noted, further, that all such boundary rays extend in the same direction on h. Thus, in Fig. IV, 10 the rays MP', $P'P$, PN are boundary rays.

An immediate consequence of Theorem 5 is that two boundary parallels cannot have a common perpendicular. Of course we proved this earlier (see

Theo. 1), but the present theorem offers an additional simple proof. Specifically, since $\angle NP'Q'$ in Figs. IV, 9 and IV, 10 is an angle of parallelism, it is acute. Hence line $P'Q'$, which is the perpendicular to g through P' (this being any point of h other than P), is not also perpendicular to h. As for line PQ, we already knew it was not perpendicular to both g and h.

We now answer another of the questions raised at the beginning of this section.

Theorem 6. *If one line is a boundary parallel to a second, then the second is a boundary parallel to the first. The relation of the directions of parallelism is such that the boundary rays on each line project into the boundary rays on the other line.*

Proof. Let g be any line (which, for convenience, we take horizontal), P any point not on it, h the right-hand parallel to g through P, and Q the projection of P on g (Fig. IV, 11). According to Theorem 4 and its proof, there

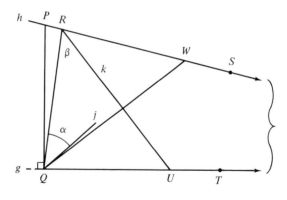

Fig. IV, 11

is a point R on h to the right of P such that $\angle RQT = \angle QRS$. We shall show that every line j subdividing $\angle RQT$ meets h. Since we already know that g does not meet h, it will then follow by Theorem 3 that g is a boundary parallel to h through Q.

Let j make an angle α with line QR. Let line k subdivide $\angle QRS$ so as to make an angle β with line QR such that $\alpha = \beta$. Line k will meet g in some point U since, by Theorem 5, h is the right-hand parallel to g through R. Take W on h to the right of R so that $RW = QU$. Then triangles RQU and QRW are congruent since $\angle RQT = \angle QRW$. It follows that $\angle RQW$ equals β, and hence α. Thus j coincides with line QW and hence meets h in W. For the reason given at the end of the preceding paragraph, then, g is a boundary parallel to h through Q. Also, g will clearly be parallel to h in the direction on h indicated by the arrow.

Finally, we verify the last sentence in the theorem by noting (1) that each ray on g in the indicated direction projects into a ray on h in the indicated direction, and vice versa, and (2) that all such rays are boundary rays.

In view of Theorem 6 we may speak of two lines g, h as being boundary parallels without saying which is parallel to which. And if a direction of parallelism on one of the lines is specified, we will know from Theorem 6 the direction of parallelism on the other. Our diagrams from now on will indicate both directions by arrows.

Before proceeding further it may be worthwhile to restate the following facts concerning two boundary parallels g, h: (1) every point on g and h is the endpoint of a boundary ray; (2) all boundary rays on g extend from their endpoints in the same direction; (3) all boundary rays on h extend from their endpoints in the same direction; (4) the boundary rays on each line project into the boundary rays on the other line.

We now answer the remaining one of the three questions raised at the beginning of this section.

Theorem 7. *If two lines are boundary parallels to a third in a given direction, then they are boundary parallels to one another, the direction of parallelism on each being the same as in its parallelism with the third line.*

Proof. Let h and j be right-hand parallels to g. Consider, first, the case in which h and j lie on different sides of g (Fig. IV, 12). Then h and j do not meet. Let A be a point on h, and B a point on j. Then segment AB meets g in some point C (III, §2, Property 1). To prove that h is parallel to j in the direction indicated on j it remains to show that every line r which subdivides

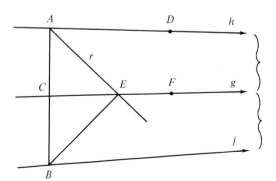

Fig. IV, 12

$\not\prec BAD$ meets j (Theo. 3), where D is on h to the right of A. Since h is a right-hand parallel to g, line r meets g in some point E (Theo. 2). Since j is a right-hand parallel to g, so g is parallel to j in the indicated direction (Theo. 6). Hence r, which subdivides $\not\prec BEF$, meets j, where F is any point on g to the right of E.

Now let h and j be on the same side of g (Fig. IV, 13). Assume h is not parallel to j in the indicated direction. In that case let k be the parallel to j in that direction through a point A on h. Then k and g lie on different sides of j and are parallel to j in the same direction. It follows from the first part of

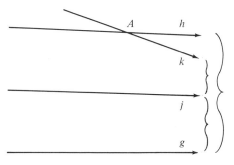

Fig. IV, 13

the proof that k is a right-hand parallel to g. Thus h and k are right-hand parallels to g through A. This contradicts that through a point not on a given line there passes a unique parallel in a specified direction. We conclude that h is parallel to j in the indicated direction.

The properties of boundary parallels stated in Theorems 5 through 7 do not involve measurement and may therefore be called *nonmetric*. We shall close this section by considering two important *metric* properties of boundary parallels.

Theorem 8.   *If two boundary parallels are met by a transversal and* α, β *are the interior angles such that one side of each is a boundary ray, then* $\alpha + \beta < 180°$.

Proof.   Let the parallels be g and h (Fig. IV, 14). If the transversal is perpendicular to one of these parallels, then one of the specified interior angles is 90° and the other is an angle of parallelism, with the result that their sum is less than 180°. Now suppose the transversal is not perpendicular to g or h. If $\alpha + \beta = 180°$, then the alternate interior angles α and γ are equal and the lines h, g have a common perpendicular (III, §7, Theo. 37), contradicting that they are boundary parallels. If $\alpha + \beta > 180°$, let line j subdivide α so that

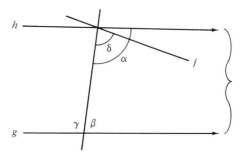

Fig. IV, 14

$\delta + \beta = 180°$. (This is always possible since $\beta < 180°$.) Then $\gamma = \delta$ and g, j are parallels with a common perpendicular. But this is impossible since j, being a subdivider of α, must meet g. It follows that $\alpha + \beta < 180°$.

We state the second metric property without proof*:

Theorem 9. *Let g, h be two boundary parallels, P any point on g, Q the projection of P on h, and f the line through P perpendicular to $\overline{PQ}$ (Fig. IV, 15). If R is any point on h in the direction of parallelism from Q, and S is its projection on f, then*

(a) *lines PQ and RS intercept equal segments $\overline{PT}$, $\overline{QR}$ on g and h;*

(b) *the angles of parallelism corresponding to R and f are equal to $\measuredangle PTS$ (that is, if the boundary parallels to f through R were drawn, each of them would make with $\overline{RS}$ an acute angle equal to $\measuredangle PTS$).*

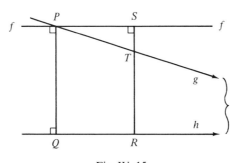

Fig. IV, 15

* For a proof, see H. E. Wolfe, *Introduction to Non-Euclidean Geometry*, p. 96 (Dryden Press, New York, 1945).

One reason for our mentioning Theorem 9 is that its part (a) leads to a simple method of constructing the boundary parallels to a line through a point not on the line (see Ex. 4).

EXERCISES

1. Let a transversal meet the boundary parallels g, h in points A, B so that α, β are the interior angles with vertices A, B, respectively, on one side of the transversal and β, γ are alternate interior angles.

(a) If all the subdividers of α meet h, what can be said about the subdividers of γ in relation to h?

(b) If all the subdividers of β meet g, what can be said about the subdividers of α in relation to h?

(c) If some subdividers of α do not meet h, what can be said about the subdividers of γ in relation to h?

(d) If all subdividers of γ meet h, which rays on g are boundary rays? which rays on h?

2. In Fig. IV, 8, where m, n, p are parallel to g in the indicated directions, we know that g is also parallel to m, n, p in certain directions. Show these directions and justify them.

3. Answer *true* or *false* to each of the following.

(a) If two lines are boundary parallels to a third, they are boundary parallels to each other.

(b) If two lines are parallel to a third, they are parallel to each other.

(c) If two boundary parallels are met by a transversal, the alternate interior angles are unequal.

(d) If two parallels are met by a transversal, the alternate interior angles are unequal.

4. Using Theorem 9, part (a), justify the following *construction of a boundary parallel* to a line h through a point P not on h. (a) Find the projection, Q, of P on h. (b) Draw line f perpendicular to $\overline{PQ}$ at P. (c) On h take any point R other than Q. (d) Find the projection, S, of R on f. (e) Draw the circle with center P and radius equal to QR. (f) If T is the point between R and S where the circle meets $\overline{RS}$, then line PT is a boundary parallel to h through P. (In justifying this construction do not merely quote Theorem 9, but give a detailed argument. You may take for granted that the circle does meet $\overline{RS}$ between R and S.)

5. In the construction of Exercise 4 show that $PS < QR < PR$ and hence that the circle can reasonably be expected to meet $\overline{RS}$ between R and S.

6. Although logically sound and easily performed, the construction in Exercise 4 is not satisfying empirically because the constructed parallel comes so close to line f as to be indistinguishable from it. Verify this and offer a possible explanation.

4. TRILATERALS

The figure consisting of two boundary parallels and a transversal (meeting them, say, in A and B) has become quite familiar to us. Now it will be useful just to deal with the part of this figure consisting of the segment $\overline{AB}$ and the two boundary rays with endpoints A, B (Fig. IV, 16). We shall call this three-sided figure a **trilateral** and refer to A, B as its *vertices*. If C, D are two points on the boundary rays, respectively, as shown in Fig. IV, 16, the trilateral can

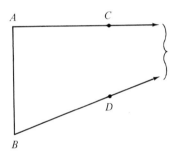

Fig. IV, 16

be designated more precisely as the *trilateral CABD*, the rays AC, BD as the *outer sides* of the trilateral, the segment $\overline{AB}$ as the *inner side* (or *middle side*), and the angles BAC and ABD as the *angles of the trilateral*. We shall say that $\angle BAC$ and side BD lie *opposite* each other, and do likewise for $\angle ABD$ and side AC. When we say that a line is a *boundary parallel to side AC*, we shall mean that it is a boundary parallel to line AC in such a way that side AC is a boundary ray. The term *boundary parallel to side BD* will have a similar meaning.

Many of the properties of trilaterals are analogous to those of triangles. The following one, for example, has already been proved and is familiar to us.

Theorem 10. *A line which subdivides an angle of a trilateral meets the opposite side.*

Another property of trilaterals recalls Pasch's Axiom for triangles:

Theorem 11. *A line which meets a side of a trilateral but does not go through a vertex will meet another side, provided that the line is not a boundary parallel to an outer side.*

Proof. Let $CABD$ be the trilateral and g the line. Suppose g meets an outer side, say AC, in P (Fig. IV, 17). For convenience let P lie between A and C. Angles α and β, as shown, are formed by $\overline{BP}$ and side AC. Line g (not shown) subdivides either α or β. If it subdivides α, it meets side $\overline{AB}$ of triangle ABP; if it subdivides β, it meets side BD of trilateral $CPBD$ (Theo. 10).

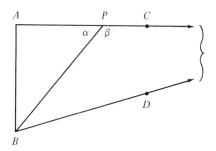

Fig. IV, 17

Thus g meets another side of trilateral $CABD$ besides AC. Now suppose g meets the inner side $\overline{AB}$ in Q (Fig. IV, 18; g not shown). Let h be the boundary parallel to side AC through Q. It is then also the boundary parallel to side BD through Q (Theo. 7). Angles γ and δ, as shown, are formed by h and $\overline{AB}$. Since g is distinct from h by hypothesis, it subdivides either γ or δ. If it subdivides γ, it meets side AC of trilateral $CAQE$ (Theo. 10); if it subdivides δ,

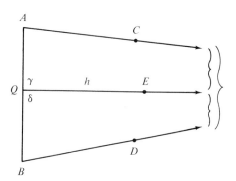

Fig. IV, 18

it meets side BD of trilateral $EQBD$. Thus g meets another side of trilateral $CABD$ besides $\overline{AB}$.

An *exterior angle* is defined for a trilateral just as for a triangle. For contrast, the angles of a trilateral may also be called its *interior angles*.

Theorem 12. *The sum of the interior angles of a trilateral is less than* $180°$. *An exterior angle is greater than the opposite interior angle.*

Proof. Consider the trilateral $CABD$ with interior angles α, β (Fig. IV, 19). By Theorem 8

$$\alpha + \beta < 180°. \tag{1}$$

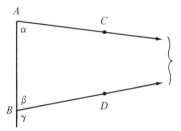

Fig. IV, 19

Consider any exterior angle of the trilateral, say the angle γ formed by extending $\overline{AB}$ in a straight line beyond B. Then

$$\beta + \gamma = 180°.$$

Because of this equation we may substitute $180° - \gamma$ for β in (1), obtaining

$$\alpha + (180° - \gamma) < 180°,$$

which reduces to $\gamma > \alpha$.

Two trilaterals are said to be *congruent* if the angles and middle side of one are equal, respectively, to the angles and middle side of the other.

Theorem 13. *Two trilaterals are congruent if an angle and the middle side of one are equal to an angle and the middle side of the other.*

Proof. Let $CABD$ and $C'A'B'D'$ (Fig. IV, 20) be trilaterals such that $\angle BAC = \angle B'A'C'$ and $AB = A'B'$. We must show that $\angle ABD = \angle A'B'D'$.

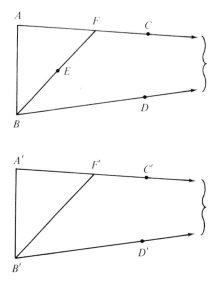

Fig. IV, 20

Assume $\angle ABD > \angle A'B'D'$. Let line BE subdivide $\angle ABD$ so that

$$\angle ABE = \angle A'B'D'. \tag{1}$$

Then line BE meets side AC in a point F. Take F' on side $A'C'$ so that $A'F' = AF$. Triangles ABF, $A'B'F'$ are congruent by side–angle–side, so that

$$\angle ABF = \angle A'B'F'. \tag{2}$$

On noting that $\angle ABF$ and $\angle ABE$ are the same angle we can combine (1) and (2), obtaining $\angle A'B'D' = \angle A'B'F'$. This is impossible since it implies that lines $B'D'$ and $B'F'$ coincide. The assumption $\angle ABD > \angle A'B'D'$ thus leads to a contradiction. The assumption $\angle ABD < \angle A'B'D'$ would do likewise. Hence $\angle ABD = \angle A'B'D'$.

Theorem 14. *Two trilaterals are congruent if the angles of one are equal, respectively, to the angles of the other.*

Proof. Let $CABD$ and $C'A'B'D'$ (Fig. IV, 21) be trilaterals such that $\angle ABD = \angle A'B'D'$ and $\angle BAC = \angle B'A'C'$. We must show that $AB = A'B'$. Assume $AB > A'B'$. Take E on $\overline{AB}$ so that $AE = A'B'$. Let line EF be the boundary parallel to side BD through E. Trilaterals $CAEF$ and $C'A'B'D'$ are then congruent (Theo. 13). Hence $\angle AEF = \angle A'B'D'$. It follows that $\angle AEF$, an exterior angle of trilateral $FEBD$, equals $\angle EBD$, the opposite interior angle

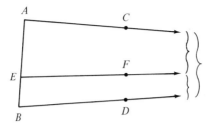

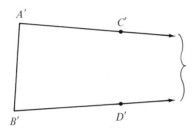

Fig. IV, 21

of this trilateral. This contradicts Theorem 12. The assumption $AB < A'B'$ leads to a like contradiction. Hence $AB = A'B'$.

EXERCISES

1. In the proof of Theorem 13 show that the assumption $\measuredangle ABD <$ $\measuredangle A'B'D'$ leads to a contradiction.

2. In the proof of Theorem 14 show that the assumption $AB < A'B'$ leads to a contradiction.

3. Prove that two equiangular trilaterals are congruent if their middle sides are equal. (A trilateral whose angles are equal is said to be *equiangular*.)

4. Consider a line which bisects the middle side of an equiangular trilateral.

(a) If the line is a boundary parallel to the other sides, show that it is perpendicular to the middle side.

(b) If the line is perpendicular to the middle side, show that it is a boundary parallel to the other sides.

5. Show that each point on the perpendicular bisector of the middle side of an equiangular trilateral is equidistant from the outer sides. (See Ex. 4.)

6. State and prove the property of isosceles triangles which is analogous to the property of equiangular trilaterals mentioned in Exercise 5.

5. ANGLES OF PARALLELISM

Thus far, all we know about the size of an angle of parallelism is that it is between 0° and 90°. Are all angles of parallelism equal? If they are not, does their size vary in some orderly way? We are now in a position to answer such questions. Let us first adopt a convenient terminology for referring to angles of parallelism. We recall that, given a line g and a point P not on it, there are two boundary parallels to g through P and two equal angles of parallelism, one on each side of the line PQ perpendicular to g at Q. These angles are called, more precisely, *the angles of parallelism corresponding to P and g.* In this view, the angles are regarded as determined by P and g inasmuch as we started out with P and g. But they may also be thought of as determined by the segment $\overline{PQ}$. For if we started out with this segment, it could be understood that P is the point through which the two parallels are to pass and that g is the perpendicular to the segment at Q. We shall therefore speak of the angles as corresponding to $\overline{PQ}$, as well as to P and g.

With the above understanding, then, to every segment $\overline{PQ}$ there correspond two equal angles of parallelism and they are acute. It is also true, and we verify this later, that every given acute angle is an angle of parallelism corresponding to some segment.* We state these important facts as a theorem.

Theorem 15. *To every given segment $\overline{PQ}$ there correspond two equal angles of parallelism and they are acute. Every given acute angle is an angle of parallelism corresponding to some segment.*

The new facts concerning the size of an angle of parallelism which we are now ready to prove are expressed in several theorems, the first of which follows.

Theorem 16. *If two segments are equal, the angles of parallelism corresponding to one are equal to the angles of parallelism corresponding to the other. Conversely, if two angles of parallelism are equal, so are the segments to which they correspond.*

Proof. Let $\overline{AB}$, $\overline{A'B'}$ be two equal segments (Fig. IV, 22). Consider the two trilaterals $CABD$ and $C'A'B'D'$ whose angles at B and B' are right angles. Then $\angle BAC$ and $\angle B'A'C'$ are angles of parallelism corresponding to $\overline{AB}$ and $\overline{A'B'}$. Since $AB = A'B'$ and the angles at B and B' are right angles, the trilaterals are congruent (Theo. 13). Hence $\angle BAC = \angle B'A'C'$. Thus the

---

* Actually, there are two segments, one on each side of the given angle.

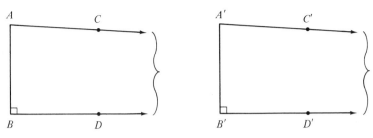

Fig. IV, 22

angles of parallelism corresponding to $\overline{AB}$ are equal to those corresponding to $\overline{A'B'}$. Conversely, let α and α' be two equal angles of parallelism. Then α corresponds to some segment $\overline{AB}$, and α' to some segment $\overline{A'B'}$ (Theo. 15). We shall suppose that $\overline{AB}$, $\overline{A'B'}$ are not the same segment (if they were, they would, of course, be equal). Consider the two trilaterals $CABD$, $C'A'B'D'$ (Fig. IV, 22) whose middle sides are $\overline{AB}$, $\overline{A'B'}$, whose angles at B, B' are right angles, and hence whose angles at A, A' are equal to α, α'. Since $\alpha = \alpha'$, these trilaterals are congruent (Theo. 14). Hence $AB = A'B'$.

Theorem 17. *To the greater of two segments there corresponds the smaller angle of parallelism, and conversely.*

Proof. Let $\overline{AB}$ exceed $\overline{A'B'}$. Consider the trilaterals $CABD$, $C'A'B'D'$ (Fig. IV, 23) in which the angles at B, B' are right angles, and hence in which the angles at A, A' are angles of parallelism corresponding to $\overline{AB}$, $\overline{A'B'}$. Take E on $\overline{AB}$ so that $BE = B'A'$. Let line EF be parallel to line BD in the indicated direction. Then trilaterals $FEBD$ and $C'A'B'D'$ are congruent (Theo. 13). Hence

$$\angle BEF = \angle B'A'C'. \tag{1}$$

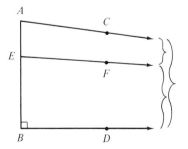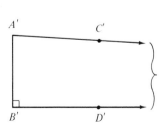

Fig. IV, 23

Noting that $\angle BEF$ is an exterior angle of trilateral $CAEF$, we infer that

$$\angle BEF > \angle BAC \qquad (2)$$

by Theorem 12. From (1) and (2) we obtain $\angle BAC < \angle B'A'C'$. This proves the first part of the theorem.

Conversely, let α and α' be angles of parallelism such that $\alpha < \alpha'$. Let $\overline{AB}$, $\overline{A'B'}$ be segments to which α, α' correspond (Theo. 15). By Theorem 16, $\overline{AB}$ and $\overline{A'B'}$ cannot be equal, for that would imply that $\alpha = \alpha'$. By the first part of the present theorem $\overline{A'B'}$ cannot exceed $\overline{AB}$, for that would imply that $\alpha' < \alpha$. It follows that $\overline{AB}$ exceeds $\overline{A'B'}$.

The preceding two theorems describe the relation between the size of a segment and the size of a corresponding angle of parallelism in a purely geometrical way since numbers are not used or involved. "Equal segments" simply means congruent segments, and they, according to Euclid, are segments that can be made to coincide. Even according to modern views the concept of congruence is usually not numerical. A similar remark holds for the term "equal angles." The term "unequal segments" means noncongruent segments, and to say that $\overline{AB}$ is greater than $\overline{A'B'}$ simply means that $\overline{A'B'}$ is congruent to a subsegment of $\overline{AB}$. The term "unequal angles" is defined in a similar manner.

However, the relation between the size of a segment and the size of a corresponding angle of parallelism can also be described numerically. To do this we choose a *unit segment* and a *unit angle*, that is, some segment and some angle in terms of which to measure all segments and angles. Then, as in Euclidean geometry, associated with every segment there is a positive number called the *length of the segment* and possessing the following properties: (1) equal segments have equal lengths; (2) the greater of two segments has the greater length; (3) the length of a segment equals the sum of the lengths of the subsegments resulting from any partition of the segment. Similarly, associated with every angle there is a positive number called the *magnitude (or measure) of the angle* and possessing three analogous properties.

Now let x and y be variables that represent lengths and angular magnitudes, respectively. Any given value of x then determines an entire set of segments, namely, those segments whose lengths have that value. Since all these segments are necessarily equal, the angles of parallelism which correspond to them are equal by Theorem 16, and hence have the same magnitude y. Thus, to any given value of x there corresponds a unique value of y. In other words, the magnitude of an angle of parallelism corresponding to a given segment is a function of the length of the segment. It is customary, following Lobachevsky, to indicate the existence of this function by writing

$$y = \Pi(x), \tag{1}$$

where the Greek capital letter Π (pi), suggested by the word "parallel," serves as the functional symbol instead of the usual f. Being the length of a segment, x may have any positive value in the function (1). The values of y in (1) are also positive, but of limited range since angles of parallelism are acute. If, for example, the angular unit is a degree, then $y < 90$.

What is the nature of the function (1)? Since, according to Theorem 17, the smaller angle of parallelism always corresponds to the greater segment, we see that (1) is a decreasing function: y always decreases as x increases. On the basis of our work nothing more can be stated about the function. It can be shown, however, by the use of ideas and methods beyond the scope of this book, that when a certain unit segment is used the precise function is

$$y = 2 \arctan(e^{-x}), \tag{2}$$

where e is the base of natural logarithms.*

Examining this equation, we verify that y does indeed decrease when x increases, approaching 0 when x becomes infinite. Also, we note that when x approaches 0, y increases and approaches 90 (if degree measure is used). In other words, when the length of a segment becomes arbitrarily great, the corresponding angle of parallelism approaches $0°$, and when the length becomes arbitrarily small, the angle approaches $90°$. A conventional Cartesian graph of equation (2) is shown in Fig. IV, 24, where degree measure is used

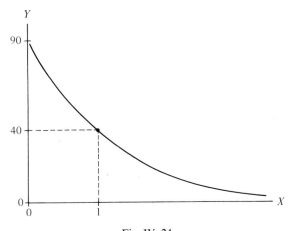

Fig. IV, 24

* The value of e to three decimal places is 2.718.

for y. The curve always falls when traversed from left to right, goes through the point $x = 1$, $y = 40$ (approximately), and is asymptotic to the X-axis. Traversed from right to left, the curve comes arbitrarily close to the point $x = 0$, $y = 90$ on the Y-axis.

Earlier we stated without proof that every given acute angle is an angle of parallelism corresponding to some segment (Theo. 15). This can now be verified with the aid of the function (2). For not only does the function determine a unique value of y between 0 and 90 (if degree measure is used) corresponding to each positive value of x, but every value of y between 0 and 90 is the correspondent of a unique value of x. This is seen by solving (2) for x,

$$x = \log_e(\cot \tfrac{1}{2}y),$$

and noting that if y is a number between 0 and 90, then $\tfrac{1}{2}y$ is between 0 and 45, $\cot \tfrac{1}{2}y$ has a value exceeding 1, and x has a value exceeding 0.

We are in no position to describe completely the special unit segment used in obtaining equation (2). This is done in the next chapter. But we can learn something about the segment at this point. Substituting 1 for x in equation (2) and using degree measure for y, we obtain

$$y = 2 \arctan(e^{-1}) = 2 \arctan 0.3679 = 40,$$

approximately. Thus the unit segment is a segment to which there corresponds an angle of parallelism of about 40°. Such a segment appears to be of very considerable extent when interpreted physically. One simple way of seeing this is to note that lines drawn perpendicular to one side of a 40° angle on a very large chart or blackboard (Fig. IV, 25) will always meet the other side, and hence not be parallel to it. This means that the unit segment, interpreted physically, is longer than the side to which the perpendiculars are drawn. To be sure, in this drawing the sides of the angle would not be more than a few

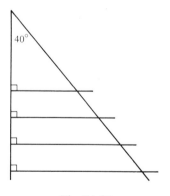

Fig. IV, 25

feet, but like results would be obtained even if the sides were thousands of miles long, for extremely large right triangles containing angles of 40° occur in astronomical work.

Despite our inability to represent the unit segment physically, it does exist mathematically, being a segment whose corresponding angle of parallelism is exactly 2 arctan(e^{-1}) degrees. A unit segment that can be so completely identified is called an *absolute unit of length*, as was mentioned in the discussion of Lambert's work (II, §6). Every segment can be identified in a similar way since each has a corresponding angle of parallelism. It follows that any segment which serves as a unit segment is an absolute unit of length and can be chosen simply by specifying its corresponding angle of parallelism. Thus we can say, for example, "let the unit segment be a segment whose corresponding angle is 80°" or "let the unit segment be a segment whose corresponding angle is 20°," and so forth. It will be recalled from the discussion of Lambert's work in Chapter II that in Euclidean geometry there is no way of identifying a segment and hence all units of length are relative, never absolute.

The merit of the special unit segment used in obtaining formula (2) is that it provides us with the simplest possible equation relating x and y. When it is not used, and when no other unit segment is specified, the equation relating x and y is

$$y = 2 \arctan(e^{-x/k}), \tag{3}$$

where k is positive and unknown. Equation (2) is seen to be the special case of (3) when $k = 1$. As this correctly suggests, when a particular value is assigned to k, the equation to which (3) reduces for that value gives the relation between x and y for some special unit segment. Thus, for $k = 2$, we obtain

$$y = 2 \arctan(e^{-x/2}). \tag{4}$$

Since $y = 2 \arctan(e^{-1/2})$ when $x = 1$, the unit segment now is a segment whose corresponding angle of parallelism is 2 arctan($e^{-1/2}$) degrees, or about 62°. In view of Theorem 17, this unit segment is shorter than the unit segment used in equation (2); that is, it is congruent to a part of the latter segment.

EXERCISES

1. Verify the graph of equation (2) shown in Fig. IV, 24 by drawing it from a table of values. Take $x = \frac{1}{2}$, 1, $\frac{3}{2}$, 2, 3, and use scales on the axes like those in the figure.

2. Find the angle of parallelism (nearest degree) corresponding to the unit segment when $k = 3$.

3. Without using tables, show that the unit segment for $k = 3$ is shorter than for $k = 2$.

4. Show that the unit segment always becomes shorter as k increases.

5. If $\overline{AB}$ is the unit segment when $k = 1$ and C is the midpoint of $\overline{AB}$, show that $\overline{AC}$ is the unit segment for $k = 2$.

6. Show that doubling k always yields a new unit segment which is half of the original.

7. Generalize Exercise 6 by showing that if k is multiplied by a positive number r ($\neq 1$), a new unit segment is obtained which is $1/r$ times the original.

8. Using the fact stated in Exercise 7, show that the angle of parallelism corresponding to a given segment, as computed by formula (3), is always the same no matter what value of k is used.

9. If m, n are any intersecting lines, show that there are four lines each of which is a boundary parallel to m and n.

6. DISTANCE BETWEEN TWO LINES

We have already learned something about the variation in distance between two parallels with a common perpendicular (III, §7, Theo. 39) and will add to that knowledge after considering the variation in distance between two intersecting lines. Finally we will study the variation in distance between two boundary parallels.

Theorem 18.   *The distance from a point on either of two intersecting lines to the other always increases, becoming arbitrarily great, as the point recedes from their point of intersection.*

Proof.   Let g, h be any intersecting lines and let them meet in A (Fig. IV, 26). Take any points P_1, P_2 on either line, say on g, so that P_1 is between A and P_2. Then P_2 is further from A than is P_1. Let Q_1, Q_2 be the projections of P_1, P_2 on h. From a consideration of triangles AP_1Q_1, AP_2Q_2 we see that the angles at P_1 and P_2 in quadrilateral $P_1Q_1Q_2P_2$ are obtuse and acute, respectively. It follows that $P_2Q_2 > P_1Q_1$ (III, §3, Theo. 33). Thus the distance from a point on g to h always increases as the point recedes from A.

To show that this distance becomes arbitrarily great we shall prove that, given any segment $\overline{CD}$, a point P can be found on g such that the distance from P to h exceeds CD. Take point R on h so that $\overline{AR}$ is a segment whose corresponding angle of parallelism is $\sphericalangle P_1AQ_1$. If j is the line through R perpendicular to h, then g is the boundary parallel to j through A in the indicated direction. Take point S on j (on the same side of h as P_1) so that

$$RS = CD. \tag{1}$$

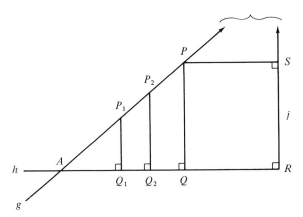

Fig. IV, 26

The perpendicular to j at S clearly cannot meet h; it must therefore meet g in some point, as is seen by considering trilateral P_1ARS (Theo. 11). Call the point P and let its projection on h be Q. Since $PQRS$ is a Lambert quadrilateral, with an acute angle at P, it follows that $PQ > RS$ (III, §6, Theo. 35). Hence, by equation (1), $PQ > CD$.

Turning now to parallels with a common perpendicular, we prove the following concerning the distance between them.

Theorem 19. *The distance from a point on any one of two parallels with a common perpendicular to the other becomes arbitrarily great as the point recedes from that perpendicular in either direction.*

Proof. Let g, h be two parallels with a common perpendicular and let it meet them in E, F (Fig. IV, 27). We already know that the distance from a point on either of the parallels to the other always increases as the point

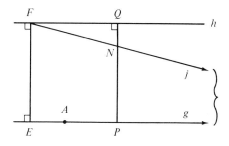

Fig. IV, 27

recedes from the common perpendicular (III, §7, Theo. 39). Now we shall prove that this distance becomes arbitrarily great. If $\overline{CD}$ is any given segment, we shall show that there is a point P on g to the right of E such that the distance from P to h exceeds CD. Let j be the right-hand parallel to g through F. Since h, j are intersecting lines, there is a point N on j such that the distance NQ from N to h exceeds CD (Theo. 18):

$$NQ > CD. \tag{1}$$

Line NQ clearly cannot meet side $\overline{EF}$ of trilateral $NFEA$, where A is any point of g to the right of E. It must therefore meet the outer side EA of this trilateral in some point P (Theo. 11). Then $PQ > NQ$. Using (1) we conclude that $PQ > CD$. By using the left-hand parallel to g through F we can show that the distance from g to h becomes arbitrarily great to the left of the common perpendicular EF. Similar results can be obtained for the distance from h to g.

We now consider the distance between boundary parallels.

Theorem 20.   *The distance from a point on either of two boundary parallels to the other always decreases as the point moves in the direction of parallelism, becoming arbitrarily small, and it becomes arbitrarily great when the point moves in the opposite direction.*

Proof.   Let g be a right-hand parallel to h (Fig. IV, 28) and let P_1, P_2 be any two points of g situated so that the direction from P_1 to P_2 is the direction of parallelism on g. If Q_1, Q_2 are the projections of P_1, P_2 on h, then the

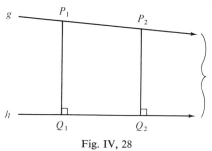

Fig. IV, 28

angles at P_1 and P_2 in quadrilateral $P_1 Q_1 Q_2 P_2$ are acute and obtuse, respectively. Hence $P_1 Q_1 > P_2 Q_2$ (III, §3, Theo. 33). Thus the distance from a point on g to h always decreases as the point moves in the direction of parallelism on g. To complete the proof we shall show that, given any segment $\overline{CD}$, there is a point on g such that the distance from the point to h equals CD.

Let P be any point of g and let Q be its projection on h. Either $PQ = CD$ or $PQ \neq CD$. In the latter case take the point R on $\overline{PQ}$ (Fig. IV, 29a), or on the extension of $\overline{PQ}$ beyond P (Fig. IV, 29b), such that

$$RQ = CD. \tag{1}$$

Let j be the left-hand parallel to h through R. Applying Theorem 11 to trilateral $MPQN$ in Fig. IV, 29a and to trilateral $MRQN$ in Fig. IV, 29b, we see that g and j must meet in a point S. Let T be the projection of S on h. Take

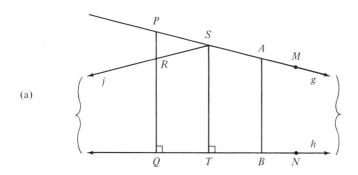

(a)

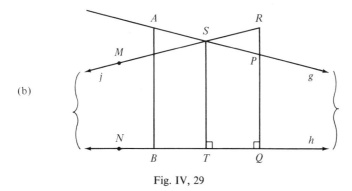

(b)

Fig. IV, 29

A on g (on the opposite side of S from P) so that $SA = SR$. On h (on the opposite side of T from Q) take B so that $TB = TQ$. From the fact that g and j are the boundary parallels to h through S, it follows that $\measuredangle AST = \measuredangle RST$. Quadrilaterals $RSTQ$ and $ASTB$ are then congruent by side–angle–side–angle–side (III, §3, Ex. 6). Hence $\overline{AB}$ is perpendicular to h and equal to $\overline{RQ}$. In other words, the distances from A and R to h are AB and RQ, and

$AB = RQ$. It follows from (1) that $AB = CD$. Thus we have exhibited a point on g, namely A, whose distance to h equals CD. Since CD is any given distance, it follows from what was shown in the first part of the proof that the distance from a point of g to h becomes arbitrarily small as the point moves in the direction of parallelism on g, and arbitrarily great as it moves in the opposite direction on g.

In this proof we considered only distances from g to h. The proof is essentially the same when distances from h to g are considered.

Inasmuch as two boundary parallels come arbitrarily close to each other in their directions of parallelism without meeting, they are asymptotic. For this reason they are often called *asymptotic parallels*, and two parallels with a common perpendicular are called *nonasymptotic parallels*.

Since two boundary parallels are asymptotic in their directions of parallelism, we may think of these directions as determining one and the same *direction in the plane*. We shall therefore speak of the two lines as being *parallel in the same direction*, meaning the same direction in the plane. If we now introduce a symbol, say a small Greek letter, to represent this common direction in the plane, we shall have a useful notation for boundary parallels. Thus, if two boundary parallels go through distinct points A and B, respectively, we may symbolize them by $A\delta$ and $B\delta$, where δ denotes the common direction in the plane determined by their directions of parallelism (Fig. IV, 30). Then $A\gamma$ and $B\gamma$ will represent two other boundary parallels through

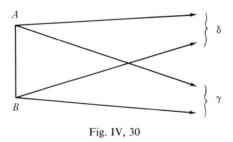

Fig. IV, 30

A and B, their common planar direction being denoted by γ. The trilateral with inner side $\overline{AB}$ and with outer sides on $A\delta$, $B\delta$ will be symbolized by $AB\delta$ or $BA\delta$.

EXERCISES

1. Prove that the distance from a point P on either of two intersecting lines to the other becomes arbitrarily small as P approaches the point of intersection. (In view of Theorem 18 it suffices to show that, given any segment, a position of P can be found such that the distance from P to the other line is less than the length of the segment.)

2. If f_1 is the figure formed by two boundary parallels, and f_2 is the figure formed by two other boundary parallels, it is intuitively clear that f_1, f_2 will be congruent if there is a trilateral with outer sides belonging to f_1 which is congruent to a trilateral with outer sides belonging to f_2. Using this as the definition of the congruence of f_1 and f_2, prove that *any two pairs of boundary parallels are congruent.*

3. Does it seem that any two pairs of nonasymptotic parallels are congruent? Justify your answer.

4. We saw (III, §14, proof of Theo. 56) that there are triangles such that the perpendicular bisectors of two sides are nonasymptotic parallels. Show that there are also triangles in which two perpendicular bisectors are asymptotic parallels.

5. Show that if the perpendicular bisectors of two sides of a triangle are asymptotic parallels, then the third perpendicular bisector is asymptotically parallel to each of them. (Use III, §14, Exs. 6, 7.)

6. If g, h are two lines meeting in A, g being horizontal, and B, C are two other points on h, give suitable notations for the asymptotic parallels to g through B, and to g through C.

7. THE UNIQUENESS OF PARALLELS WITHOUT A COMMON PERPENDICULAR

We are now in a position to complete the proof of Theorem 1 in Section 2. The theorem asserts that the boundary parallels m, n to a line g through a point F are the *only* parallels to g through F which have no common perpendicular with g. We proved that m, n have no common perpendicular with g and it remains to justify the word "only."

Let E be the projection of F on g and let h be the line perpendicular to $\overline{EF}$ at F (Fig. IV, 31). Assume that j is a parallel to g through F, different from m and n, which has no common perpendicular with g. Then j subdivides either the acute angle formed by m and h or the acute angle formed by n and h. Suppose the former position occurs. Since g, j have no common perpendicular, if a point traverses j in one direction, its distance from g either always increases or always decreases, depending on the direction (III, §7, Ex. 11). To determine the direction of decrease, take P and D on j so that F is between them, as shown, and let C be the projection of D on g. Since $\angle PFE$ is acute, $\angle DFE$ must be obtuse. Hence the angle at D in quadrilateral $CEFD$ is acute (III, §8, Theo. 44), so that $EF < CD$ (III, §3, Theo. 33). Thus, if a point

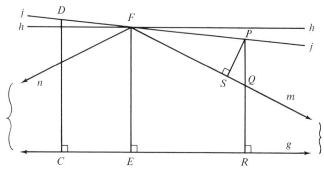

Fig. IV, 31

traverses j in the downward direction, that is, in the direction from D to F, its distance to g always decreases.

Now let R be the projection of P on g and let Q be the point in which $\overline{PR}$ meets m. Then

$$PQ = PR - QR. \tag{1}$$

If S is the projection of P on m, then QPS is a right triangle and

$$PS < PQ. \tag{2}$$

Combining (1) and (2) gives

$$PS < PR - QR,$$

or

$$PR > PS + QR. \tag{3}$$

Let P traverse j in the downward direction. Then, since j and m are intersecting lines, the distance PS increases and becomes arbitrarily great; and, since g, m are boundary parallels, the distance QR decreases, becoming arbitrarily small. Thus $PS + QR$ becomes arbitrarily great. According to (3), PR must do likewise. But this is inconsistent with the fact (shown earlier in the proof) that the distance from j to g always decreases in the downward direction on j. From this contradiction we conclude that a line such as j cannot exist.

It is clear that the proof would be essentially the same if j subdivided the angle formed by h and n.

EXERCISES

1. Answer *true* or *false* to each of the following statements, where a, b denote two lines.

(a) If a, b have a common perpendicular, they are parallel.

(b)   If a, b are parallel, they have a common perpendicular.

(c)   If a, b have no common perpendicular, they are boundary parallels.

(d)   If a, b are boundary parallels, they have no common perpendicular.

(e)   If a, b are not boundary parallels, they have a common perpendicular.

(f)   If a point moves along line a in one direction and, as it does so, its distance to line b always decreases, then a and b are asymptotic parallels.

8.   PERPENDICULAR BISECTORS OF THE SIDES OF A TRIANGLE

We have already noted the following two facts concerning the perpendicular bisectors of the sides of a triangle:

1.   *If two perpendicular bisectors of the sides of a triangle meet, then the third goes through their point of intersection and the three bisectors are therefore concurrent* (III, §14, Ex. 6).

2.   *If two perpendicular bisectors of the sides of a triangle are parallel, with a common perpendicular, then all three bisectors are parallel, with this same common perpendicular* (III, §14, Ex. 7).

Also, we have noted that each of these cases actually occurs (III, §14, Ex. 8 and proof of Theo. 56). The remaining possibility is that some of the bisectors might be boundary parallels. This case occurs, too (§6, Ex. 4), and the statement for it is as follows:

3.   *If two perpendicular bisectors of the sides of a triangle are boundary parallels, then the third bisector is a boundary parallel to each of them.*

This statement is an immediate consequence of statements 1 and 2 above. We leave the proof as Exercise 1.

It will be noted that statement 3 says nothing about directions of parallelism. If, for example, the two given bisectors g, h are parallel in the direction α, statement 3 does not say whether the third bisector i is parallel to g and h in this same direction (Fig. IV. 32) or whether it is parallel to them in different directions β and γ (Fig. IV, 33). There are no other possibilities. We

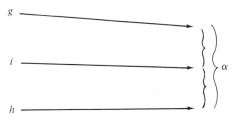

Fig. IV, 32

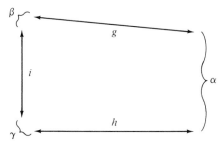

Fig. IV, 33

shall now prove that the situation in Fig. IV, 33 cannot occur, and hence that

4.   *If the three perpendicular bisectors of the sides of a triangle are boundary parallels to each other, they are all parallel in the same direction.*

We start the proof by establishing a simple fact about the perpendicular bisectors for any triangle: There is always a line which intersects all three bisectors. Let ABC be any triangle (Fig. IV, 34). If one of its angles exceeds each of the others, suppose that one is $\angle C$. Then $\angle C$ can be subdivided so that we obtain a triangle ACD in which $\angle CAD = \angle ACD$, where D is some

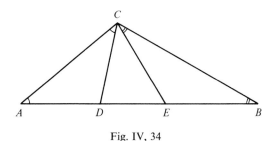

Fig. IV, 34

point on $\overline{AB}$. Similarly, triangle BCE can be formed in which $\angle CBE = \angle BCE$, where E is on $\overline{AB}$. Clearly, each perpendicular bisector for triangle ABC meets line AB. This same conclusion is also easily reached if $\angle C$ exceeds $\angle A$ but equals $\angle B$, or if the three angles are equal. We leave the details as Exercise 2.

To complete the proof of statement 4 we need only show that there is no line which meets three lines g, h, i which are mutually parallel as in Fig. IV, 33, for it would then follow that g, h, i cannot be the perpendicular bisectors for a triangle. Using Fig. IV, 35, which is essentially Fig. IV, 33 enlarged, let us consider a line m which meets g and i in A and B, and show that it cannot also meet h. Let j be the line through A parallel to h in the direction γ. Then g and

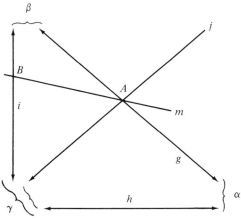

Fig. IV, 35

j are the two boundary parallels to i through A. Since m goes through A and meets i, it subdivides a certain pair of vertical angles formed by g and j (Theo. 1). If m met h, it would have to subdivide the other pair too, for g and j are also the boundary parallels to h through A. This being impossible, m cannot meet h. Statement 4 is thus proved.

Combining statements 1 through 4, we obtain

Theorem 21.   *The three perpendicular bisectors of the sides of a triangle are either concurrent, parallel with a common perpendicular, or parallel in the same direction.*

EXERCISES

1.   Deduce statement 3 from statements 1 and 2.

2.   Prove that there is a line which intersects the perpendicular bisectors for any triangle ABC in which (a) $\angle C$ exceeds $\angle A$ but equals $\angle B$, or (b) the three angles are equal.

V

Horocycles

1. INTRODUCTION

Other than straight lines, circles are the only curves we have studied thus far.* Our interest in circles lay mainly in comparing their properties with those of Euclidean circles. Of far greater importance in the further development of our subject, however, is another kind of curve, not found in Euclidean geometry, called a *horocycle*. This curve is basic in obtaining the formulas of hyperbolic trigonometry, that is, the numerical relations among the sides and angles of a triangle. This is done in the next chapter. In the present chapter we study horocycles themselves and it will be seen that they have considerable intrinsic interest.

2. CORRESPONDING POINTS

The points on two boundary parallels can be paired in a certain simple way which will lead us to the definition of a horocycle. Consider any two boundary parallels, say with the common direction δ, and let P be a point on one of them (Fig. V, 1).† If there is a point Q on the other such that trilateral $PQ\delta$ is equiangular, we shall say that Q *corresponds* to P. Clearly, according to this definition, if Q corresponds to P, then P corresponds to Q. We may therefore call P and Q a pair of (mutually) *corresponding points*. The

---

* Some attention also was given to *equidistant curves*, but only in exercises.
† From now on, the common direction of boundary parallels will usually be indicated by δ, δ', and so on, instead of by arrows.

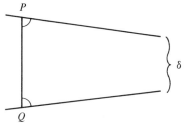

Fig. V, 1

following useful test for corresponding points can be noted immediately. Its proof is left as an exercise.

Theorem 1. *If points P and Q lie, respectively, on two lines parallel in the direction δ, they are corresponding points on these lines if and only if the perpendicular bisector of $\overline{PQ}$ is parallel to the lines in the direction δ.*

Intuition strongly suggests that if a point on one of two boundary parallels is given, then a corresponding point on the other always exists. We shall show this to be correct, but first the following must be proved.

Theorem 2. *Given any two boundary parallels, there exists a straight line each of whose points is equidistant from them. The line is parallel to them in their common direction.*

Proof. Let g, h be two boundary parallels with common direction $δ$, and A, B any two points on them, respectively (Fig. V, 2). The bisector of $\measuredangle A$ of trilateral $ABδ$ meets side $Bδ$ in a point X (not shown in Fig. V, 2)

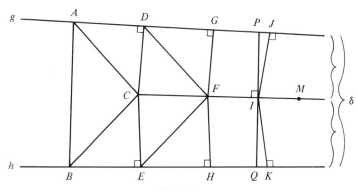

Fig. V, 2

and the bisector of $\star B$ of the trilateral meets side AX of triangle ABX in a point C. Thus the bisectors of the angles of trilateral $AB\delta$ meet in C. The perpendiculars $\overline{CD}$, $\overline{CE}$ from C to g and h are equal since each of them is equal to the perpendicular from C to $\overline{AB}$. It follows that trilaterals $CD\delta$, $CE\delta$ are congruent (IV, §4, Theo. 13), and hence that their angles at C are equal. Now consider line $C\delta$ and let F be any point on it other than C. Then triangles CDF, CEF are congruent by side–angle–side. From this we infer the congruence of triangles DFG, EFH, where G, H are the projections of F on g and h. Hence $FG = FH$. Thus every point of line $C\delta$ is equidistant from g and h.

We shall call a line each of whose points is equidistant from two boundary parallels an *equidistant line* for the two parallels.

We are now ready to prove

Theorem 3.   *Given any point on one of two boundary parallels, there is a unique point on the other which corresponds to it.*

Proof.   Let g, h be boundary parallels, and let P be any point on either of them, say on g. We shall continue to use Fig. V, 2, in which line $C\delta$ is the equidistant line for g and h whose existence was proved in Theorem 2. If I is the projection of P on $C\delta$, let IJ and IK be the equal distances from I to g and h. Take Q on h to the left of K so that $KQ = JP$. Right triangles IJP, IKQ are then congruent, so that

$$\star IPJ = \star IQK \tag{1}$$

and $\star PIJ = \star QIK$. From the latter equality and the fact that $\star MIJ = \star MIK$, where M is any point on $C\delta$ to the right of I, it follows that $\star QIM$ is a right angle since $\star PIM$ is. Thus P, I, Q are collinear. From this and (1) we see that $PQ\delta$ is an equiangular trilateral, and hence that Q corresponds to P. It is left as Exercise 3 to show that there can be no other point on h which corresponds to P.

It should be noted in the preceding proof that $\overline{PQ}$ is bisected at right angles by line $C\delta$. Thus Q is symmetrical to P with respect to this line. Line $C\delta$ is therefore not only an equidistant line for g and h, but also an axis of symmetry for them, each two corresponding points on g, h being symmetrical to this axis. Before stating a theorem to express this fact, let us consider a question we have thus far avoided. We know that each point on line $C\delta$ is equidistant from g and h. Could there be a point equidistant from them which is not on $C\delta$? Assume such a point R exists. It is easily shown that R must lie in the region between g and h (Ex. 2). Consider line $R\delta$, which, of course, is distinct from line $C\delta$. Since R is equidistant from g and h, we can prove, just as we did for line $C\delta$, that line $R\delta$ is an equidistant line for g, h and that each two corresponding points P, Q on g, h are symmetrical to line $R\delta$. Hence line PQ is perpendicular to line $R\delta$ as well as to line $C\delta$. This being impossible, a point such as R cannot exist. We have therefore proved

Theorem 4. *The locus of points equidistant from any two boundary parallels g, h is a straight line parallel to them in their common direction and situated in the region between them. This line is an axis of symmetry for g, h, their only such axis, and each two corresponding points on g, h are symmetrical to it.*

We next prove

Theorem 5. *If three points P, Q, R lie, respectively, on three parallels in the same direction so that P, Q are corresponding points on their parallels and Q, R are corresponding points on theirs, then P, Q, R are noncollinear.*

Proof. Let the direction be δ and suppose, first, that line $Q\delta$ is between the others* (Fig. V, 3). Since P, Q and Q, R are pairs of corresponding

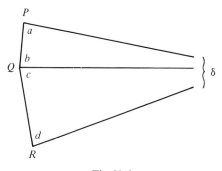

Fig. V, 3

points, each of the trilaterals $PQ\delta$, $QR\delta$ is equiangular:

$$a = b \quad \text{and} \quad c = d.$$

Also (IV, §3, Theo. 8),

$$a + b < 180° \quad \text{and} \quad c + d < 180°.$$

* That is, the others are on different sides of line $Q\delta$.

It follows that a, b, c, d are acute. Hence $b + c < 180°$, which shows that P, Q, R cannot be collinear. The proof of the theorem for the case in which line $Q\delta$ is not between the others is left as an exercise.

Theorem 6. *If three points P, Q, R lie, respectively, on three parallels in the same direction so that P, Q are corresponding points on their parallels and Q, R are corresponding points on theirs, then P, R are corresponding points on their parallels.*

Proof. Let the direction be δ and suppose that line $Q\delta$ is between the others (Fig. V, 3). Since P, Q are corresponding points, the perpendicular bisector of $\overline{PQ}$ is parallel to lines $P\delta$, $Q\delta$ in the direction δ (Theo. 1). Similarly, the perpendicular bisector of $\overline{QR}$ is parallel to lines $Q\delta$, $R\delta$ in the direction δ. Thus the perpendicular bisectors of $\overline{PQ}$ and $\overline{QR}$ are parallel in the direction δ. Now, P, Q, R are the vertices of a triangle by Theorem 5, and $\overline{PQ}$, $\overline{QR}$ are two sides of this triangle. Hence the perpendicular bisector of $\overline{PR}$, the third side, is parallel to the perpendicular bisectors of $\overline{PQ}$ and $\overline{QR}$ in the direction δ (IV, §8, Theo. 21). It is therefore also parallel to lines $P\delta$ and $R\delta$ in the direction δ. By Theorem 1, then, P and R are corresponding points on $P\delta$ and $R\delta$. The proof for the case in which line $Q\delta$ is not between the others is left as an exercise.

EXERCISES

1. Prove Theorem 1.

2. If a point is equidistant from two boundary parallels, show that it must lie in the region between them.

3. Complete the proof of Theorem 3 by showing that Q is the only point on h which corresponds to P.

4. Prove that the equidistant line for two boundary parallels is their only axis of symmetry by showing, without using Theorem 4, that if a line is an axis of symmetry for two boundary parallels, then it is also their equidistant line.

5. Show that any two given points are corresponding points on (a) at least one pair of boundary parallels, (b) exactly two pairs of boundary parallels.

6. Prove Theorem 5 for the case in which line $Q\delta$ is not between the others.

7. Prove Theorem 6 for the case in which line $Q\delta$ is not between the others.

3. DEFINITION OF A HOROCYCLE

Consider any line g, any point P on it, and a direction δ on g (Fig. V, 4). On each line parallel to g in the direction δ there is a unique point Q that corresponds to P. The locus consisting of P and all such points Q is called a **horocycle**, or, more precisely, the horocycle determined by g, P, and δ. The lines parallel to g in the direction δ, together with g, are called the *radii* of the horocycle. Since g can be denoted by $P\delta$, we may regard the horocycle as determined simply by P and δ, and hence call it the *horocycle through P with direction δ*, or, in symbols, the horocycle (P, δ).

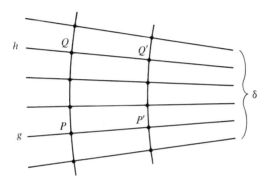

Fig. V, 4

All the points of this horocycle being mutually corresponding points by Theorem 6, the horocycle is equally well determined by any one of them and δ. In other words, if Q is any point of horocycle (P, δ) other than P, then horocycle (Q, δ) is the same as horocycle (P, δ). If P' is any point of g other than P, then horocycle (P', δ) is clearly different from horocycle (P, δ), although they have the same direction and the same radii (Fig. V, 4). We shall call distinct horocycles which have the same direction, and hence the same radii, *codirectional horocycles*.

As the terms "cycle" and "radii" might suggest, there are analogies between horocycles and circles. Some of them are stated later as theorems. A few others can be mentioned immediately. There is a unique circle, with a given center, which passes through a given point; there is a unique horocycle with a given direction, which passes through a given point. Two concentric circles have no common point; two codirectional horocycles have no common point. A unique radius is associated with each point of a circle; a unique radius is associated with each point of a horocycle. A *tangent to a horocycle* at a point on the horocycle is defined to be the line through the point which is perpendicular to the radius associated with the point. We shall presently see that a tangent to a horocycle meets it only in the point of tangency and that no line can meet a horocycle in more than two points.

The last-mentioned property is a consequence of the fact that no three points of a horocycle are collinear inasmuch as it is a locus of mutually corresponding points (see Theo. 5). That a tangent to a horocycle meets it only in the point of tangency is also easily seen, for let A be the point, t the tangent, and δ the direction of the horocycle (Fig. V, 5). If t met the horocycle

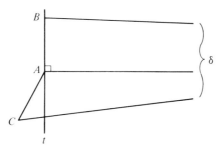

Fig. V, 5

in another point, B, we would have an equiangular trilateral $AB\delta$ with two right angles, which is impossible. In fact, the entire horocycle, except for A, lies on the same side of t, namely, the side containing the ray $A\delta$. For if a point C of the horocycle were on the other side of t, we would have an equiangular trilateral $AC\delta$ with two obtuse angles, which is impossible.

Finally, let us show that each line g (Fig. V, 6) through A other than the

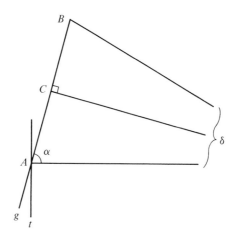

Fig. V, 6

tangent or radius meets the horocycle in one further point. (Such lines are called *secants*.) Let α be the acute angle between g and ray $A\delta$. Let C be the point of g, on the side of t containing the horocycle, such that $\overline{AC}$ is a segment corresponding to α as angle of parallelism. The line perpendicular to g at C is then parallel to line $A\delta$ in the direction δ. Let B be the point of g such that C is the midpoint of $\overline{AB}$. Trilaterals $AC\delta$, $BC\delta$ are then congruent since each contains a right angle and $AC = BC$. Hence $\sphericalangle CB\delta = \alpha$. Thus B corresponds to A and is therefore on the horocycle.

We have now completely proved

Theorem 7. *The tangent at any point A of a horocycle meets the horocycle only in A. Every other line through A except the radius meets the horocycle in one further point B. If α is the acute angle between this line and the radius, then $\overline{AB}$ is twice the segment which corresponds to α as angle of parallelism.*

EXERCISES

1. Show that a horocycle is symmetrical to each of its radii.

2. Given any horocycle, show that there are lines which do not meet it.

3. Show that there is a horocycle through any two points. Is there only one? Justify your answer.

4. Is there always a horocycle through three noncollinear points? Explain.

4. ARCS AND CHORDS OF A HOROCYCLE

By a *chord* of a horocycle we shall mean, of course, the segment joining two points of the horocycle. The proofs of the following two theorems are left to the student.

Theorem 8. *The straight line which bisects a chord of a horocycle at right angles is a radius of the horocycle.*

Theorem 9. *If $\overline{AB}$, $\overline{A'B'}$ are equal chords of horocycles with directions δ, δ', the trilaterals $AB\delta$ and $A'B'\delta'$ are congruent. (The horocycles need not be distinct.)*

The discussion that follows will prepare the ground for defining an arc of a horocycle. Consider two lines g, h parallel in the direction δ (Fig. V, 7). A third line j parallel to them in this direction will lie in one of the three regions of the plane determined by g and h (III, §2, Property 2). When j lies in the region between g and h, numbered (2) in Fig. V, 7, we shall say that it is *between* g and h.* Clearly, j will be between g and h if and only if it contains an inner point of $\overline{MN}$, where M and N are any points of g and h.

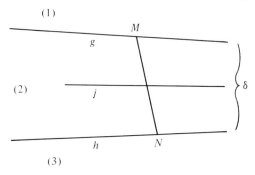

Fig. V, 7

Now let A, B be any two points of a horocycle with direction δ (Fig. V, 8). A third point of the horocycle will be said to lie *between A and B on the horocycle* if and only if it is on a radius which is between lines $A\delta$ and $B\delta$. Since each line through an inner point of $\overline{AB}$ parallel to $A\delta$ in the direction δ is such a radius, there are infinitely many points of the horocycle between A and B. The set consisting of all such between-points, together with A and B, is called the **arc AB** (in symbols, $\overset{\frown}{AB}$). Points A and B are the *endpoints* of the arc; the other points of the arc are its *interior points* (or *inner points*). The arc $\overset{\frown}{AB}$ and chord $\overline{AB}$ are said to *subtend* each other. Let C, D be any

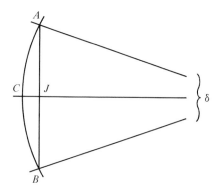

Fig. V, 8

_____
* As mentioned earlier, g and h then lie on different sides of j.

two points of $\overset{\frown}{AB}$. Then there is an arc $\overset{\frown}{CD}$, and each of its points is also a point of $\overset{\frown}{AB}$ inasmuch as each line between lines $C\delta$, $D\delta$ is also between lines $A\delta$, $B\delta$. If $\overset{\frown}{CD}$ is not identical with $\overset{\frown}{AB}$ we shall call it a *subarc* of $\overset{\frown}{AB}$.

Figure V, 8 shows an interior point C of $\overset{\frown}{AB}$. Its position to the left of line AB rather than to the right of it should be noted. To state this more precisely, if J is the point in which the radius $C\delta$ meets $\overline{AB}$, then the direction from C to J is the direction δ. As a result, $\measuredangle ACB$ is concave in the direction δ and convex in the opposite direction. Our subsequent drawings of horocycles will take this into account by showing them to be concave in their directions of parallelism. It will be left as an exercise to show that $\measuredangle ACB$ is the largest angle in triangle ABC and hence that $\overline{AB}$ is greater than each of the segments $\overline{AC}$ and $\overline{BC}$. In other words, the chord subtending $\overset{\frown}{AB}$ is greater than the chord subtending subarc $\overset{\frown}{AC}$ or subarc $\overset{\frown}{BC}$. From this one can easily obtain the more general result:

Theorem 10. *The chord subtending an arc of a horocycle exceeds the chord subtending any subarc, or, stated differently, the distance between the endpoints of an arc exceeds the distance between any other two points of the arc.*

In order to compare arcs of horocycles with each other we shall need to know when two such arcs are congruent. Since an arc is a set of points, the following definition will be used for that purpose. Two sets of points are called *congruent* if and only if there exists a one-to-one pairing* of their points such that the distance between each two points in one set equals the distance between the two points in the other set with which they are paired. Such a one-to-one pairing of the points of two sets is said to be *distance-preserving*. It is therefore often called an *isometry*. We may then say that congruent sets are sets between which an isometry exists and noncongruent sets are sets between which no isometry exists. It can be shown that this new concept of congruence is consistent with the one we have used previously. For example, if two triangles are congruent in the familiar sense of having their sides and angles equal, respectively, then they are also congruent in the sense that there is an isometry between their points.

The following theorem can now be proved.

---

* That is, any point A in either set is paired with a unique point A' in the other set. A one-to-one pairing is usually called a *one-to-one correspondence* or a *one-to-one mapping* and each of the points A, A' is said to be *mapped* on the other.

Theorem 11. *The arcs subtending two chords of the same or different horocycles are congruent if and only if the chords are equal.*

Proof. Let $\overline{AB}$, $\overline{A'B'}$ be two chords, respectively, of horocycles h, h' with directions δ, δ'. Consider, first, the case in which the chords are equal (Fig. V, 9). (For convenience in drawing the diagram we have taken h, h' to be different, but it will be seen that the proof holds even if they are the same.)

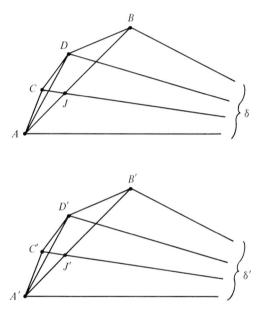

Fig. V, 9

To show that there is an isometry between the sets $\overset{\frown}{AB}$, $\overset{\frown}{A'B'}$, we must first show how to pair their points. Let us pair A' with A, B' with B, and seek the point to be paired with an interior point C of $\overset{\frown}{AB}$. Since C is an interior point of $\overset{\frown}{AB}$, radius $C\delta$ meets $\overline{AB}$ in a point J. Take J' on $\overline{A'B'}$ so that $A'J' = AJ$. Denote by C' the point on line $J'\delta'$ such that J' is on ray $C'\delta'$ and $C'J' = CJ$.

We pair C' with C and go on to show that C' is an interior point of $\overset{\frown}{A'B'}$. Trilaterals $AB\delta$, $A'B'\delta'$ being congruent (Theo. 9), so are trilaterals $AJ\delta$, $A'J'\delta'$, and hence also triangles ACJ, $A'C'J'$. The latter congruence implies the congruence of trilaterals $AC\delta$, $A'C'\delta'$. Since A, C are points of horocycle h, trilateral $AC\delta$ is equiangular. Hence trilateral $A'C'\delta'$ is also equiangular. Thus C' is a point of horocycle h'. Inasmuch as radius $C'\delta'$ intersects $\overline{A'B'}$, C' is by definition an interior point of $\overset{\frown}{A'B'}$.

Starting out with a definite point C of $\overset{\frown}{AB}$ we have exhibited a method of determining a definite point C' of $\overset{\frown}{A'B'}$ to pair with it. Had we started with this same C' and applied the method in reverse, we would clearly have determined the same C. In other words, having shown how to find the point of $\overset{\frown}{A'B'}$ to pair with each given point of $\overset{\frown}{AB}$, we also know that this pairing procedure accounts for *every* point of $\overset{\frown}{A'B'}$; that is, each point of $\overset{\frown}{A'B'}$ has been paired with some point of $\overset{\frown}{AB}$. Thus there is a one-to-one pairing of the points of $\overset{\frown}{AB}$ and $\overset{\frown}{A'B'}$. To show that the pairing preserves distance we note, first, that $AB = A'B'$, $AC = A'C'$, and $BC = B'C'$. Similarly, if D, D' are any other paired points, then $AD = A'D'$ and $BD = B'D'$. Finally, from the congruence of triangles ACD, $A'C'D'$ by side–angle–side it follows that $CD = C'D'$. Thus distance is preserved and there is an isometry between the points of $\overset{\frown}{AB}$ and $\overset{\frown}{A'B'}$. These arcs are therefore congruent.

It remains to consider the case in which chords $\overline{AB}$, $\overline{A'B'}$ are unequal. One of them being the greater, suppose it is $\overline{AB}$. Then

$$AB > A'B'. \tag{1}$$

If $\overset{\frown}{AB}$, $\overset{\frown}{A'B'}$ were congruent, there would be an isometry between their points. In particular, the points A, B on $\overset{\frown}{AB}$ would be paired with two points X', Y' on $\overset{\frown}{A'B'}$ such that $AB = X'Y'$. Hence, by (1), $X'Y' > A'B'$. This contradicts that the distance between two points of a horocyclic arc cannot exceed the distance between the endpoints of the arc (Theo. 10). It follows that $\overset{\frown}{AB}$, $\overset{\frown}{A'B'}$ are not congruent.

The terms "equal" and "congruent" are often used interchangeably in geometry. We say "all right angles are equal" and also "all right angles are congruent." In the preceding discussion we spoke of "equal chords" but could have said "congruent chords" instead. To make the terminology of Theorem 11 more symmetrical, and also to give the statement of the theorem a more familiar ring, let us agree to speak of congruent arcs as *equal arcs*.* The theorem can then be stated: *The arcs subtending two chords of the same or different horocycles are equal if and only if the chords are equal* or, more familiarly, *equal chords subtend equal arcs and equal arcs subtend equal chords.*

---

* It should be noted that we say nothing about the *length* of an arc, which is a number. This term is considered later and we shall see that if two arcs are equal (that is, congruent), then they have the same length.

It follows from Theorem 11 that unequal chords of the same or different horocycles subtend unequal (that is, noncongruent) arcs, and vice versa. If $\overparen{AB}$, $\overparen{A'B'}$ are any such arcs, we shall say that $\overparen{A'B'}$ is *less than* $\overparen{AB}$, or that $\overparen{AB}$ is *greater than* $\overparen{A'B'}$, if and only if $\overparen{A'B'}$ is equal to a subarc of $\overparen{AB}$. The following theorem then holds. We leave its proof as an exercise.

Theorem 12.   *The greater of two chords of the same or different horocycles subtends the greater arc, and vice versa.*

Let $\overparen{AB}$ be any arc of a horocycle (Fig. V, 10). The line which is perpendicular to $\overline{AB}$ and goes through its midpoint D is a radius of the horocycle (Theo. 8) and hence contains a point C of the horocycle. This point C is clearly an interior point of $\overparen{AB}$. Also, it divides $\overparen{AB}$ into equal arcs $\overparen{AC}$, $\overparen{BC}$

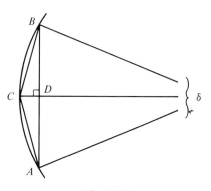

Fig. V, 10

inasmuch as the chords $\overline{AC}$, $\overline{BC}$ are equal. No other point of $\overparen{AB}$ can divide it into equal arcs. (For, the chords subtending these arcs being equal, the line joining such a point to D would have to be perpendicular to $\overline{AB}$ and yet be different from line CD, which is impossible.) We therefore call C the *midpoint* of $\overparen{AB}$.

Theorem 13.   *There is a unique point on any arc of a horocycle that divides the arc into two equal subarcs. This point, called the midpoint of the arc, is on the radius which bisects at right angles the chord subtending the arc.*

We are now in a position to obtain an important result by considering a special case of Theorem 7. According to this theorem, a line which goes through any point A (Fig. V, 11) of a horocycle with direction δ and makes a 45° angle with the radius $A\delta$ will meet the horocycle again in a point B such that $\overline{AB}$ is twice the segment corresponding to an angle of parallelism of 45°. In other words, if C is the midpoint of $\overline{AB}$, then $\overline{AC}$ is a segment whose corresponding angle of parallelism is 45°. Since t, the tangent at A, is perpendicular to the radius $A\delta$, the acute angle between t and line AB is 45°.

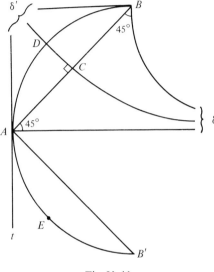

Fig. V, 11

Noting that $\overline{AC}$ is perpendicular to line $C\delta$,* we infer that t is parallel to line $C\delta$ in the direction on this line *opposite* to δ. The same can be said for the tangent at B since $BC = AC$ and $\not\prec CB\delta = 45°$. We state this result as follows.

Theorem 14. *If a chord of a horocycle is twice the segment corresponding to a 45° angle of parallelism, then the tangent at each endpoint of the chord is parallel to the perpendicular bisector of the chord in the direction on the bisector which is opposite to that of the horocycle.*

It is useful to have a name for the chords dealt with in Theorem 14. Since they are the chords which make angles of 45° with the radii through their endpoints and are all equal, we shall call them *45°-chords*. From previous

* It was helpful in drawing Fig. V, 11 to distort lines $B\delta$ and $C\delta$.

discussions it is clear that each point A of a horocycle is an endpoint of just two 45°-chords, the other endpoints B, B' of these chords being symmetrical to the radius through A (Fig. V, 11).

The perpendicular bisector of a 45°-chord divides the subtended arc into two equal arcs, which (for lack of a suitable name) we shall call *k-arcs*. All such subtended arcs being equal, so are all k-arcs equal. In Fig. V, 11, for example, where the radius $C\delta$ bisects the 45°-chord $\overline{AB}$ at right angles, if D is the point of the horocycle which is on $C\delta$, then $\overarc{AD}$ and $\overarc{BD}$ are k-arcs. Each point of a horocycle is clearly an endpoint of just two k-arcs. Point A in Fig. V, 11, for example, is an endpoint of the k-arcs $\overarc{AD}$ and $\overarc{AE}$, where E is the midpoint of $\overarc{A'B'}$. It will be useful to restate Theorem 14 in terms of k-arcs:

Theorem 15.   *The tangent at an endpoint of a k-arc is parallel to the radius through the other endpoint in the direction along the radius which is opposite to that of the horocycle.*

EXERCISES

1.   Prove Theorem 8.

2.   Prove Theorem 9.

3.   If C is an interior point of a horocyclic arc $\overarc{AB}$, prove that $\measuredangle ACB$ is the largest angle in triangle ABC, and hence that $\overline{AB}$ is greater than each of the segments $\overline{AC}$ and $\overline{BC}$.

4.   Prove Theorem 10.

5.   Prove Theorem 12. (Assume $\overline{AB}$ exceeds $\overline{A'B'}$. Using Theorem 10 and Exercise 7, show that there is an interior point C on $\overarc{AB}$ such that $AC = A'B'$. Since $\overarc{AC}$, $\overarc{A'B'}$ are then equal, $\overarc{AB}$ exceeds $\overarc{A'B'}$ by definition. The proof of the converse is straightforward.)

6.   Prove that any two horocycles are congruent.

7.   If A, B, C are distinct points on a horocycle, show that exactly one of them is between the others.

8.   Prove the following for horocyclic arcs: (a) any arc is congruent to itself; (b) if $\overarc{AB}$, $\overarc{A'B'}$ are congruent, and likewise $\overarc{A'B'}$, $\overarc{A''B''}$, then $\overarc{AB}$, $\overarc{A''B''}$ are congruent.

9. Prove the following for horocyclic arcs: If $\overset{\frown}{AB} < \overset{\frown}{A'B'}$ and $\overset{\frown}{A'B'} < \overset{\frown}{A''B''}$, then $\overset{\frown}{AB} < \overset{\frown}{A''B''}$. (Thus the relation "less than" is *transitive*. Similarly, the relation "greater than" is transitive.)

10. If two triangles are congruent in the sense that their sides and angles are equal, respectively, show that there is an isometry between their points.

11. Show that there is an isometry between the points of two circles with equal radii.

12. Find the length of a 45°-chord.

5. CODIRECTIONAL HOROCYCLES

This term, we recall from Section 3, refers to horocycles which have the same direction and hence the same radii. Their properties, which have certain analogies with those of concentric circles, are very important and will concern us for the remainder of the chapter.

By *corresponding arcs* of two codirectional horocycles we shall mean any two of their arcs whose endpoints are on the same two radii. Figure V, 12

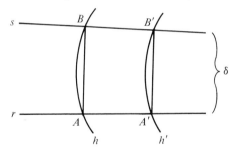

Fig. V, 12

shows corresponding arcs $\overset{\frown}{AB}$, $\overset{\frown}{A'B'}$ of codirectional horocycles h, h'; the endpoints of each arc are on the radii r and s. *Corresponding chords* of two codirectional horocycles are two chords which subtend a pair of corresponding arcs of the horocycles. In Fig. V, 12, $\overline{AB}$ and $\overline{A'B'}$ are corresponding chords of h and h'.

Since A, B belong to h, they are corresponding points and lie on the boundary parallels r and s. Similarly, A', B' are corresponding points belonging to h' and also lie on r and s. By Theorem 4, then, the axis of symmetry of r and s bisects $\overline{AB}$ and $\overline{A'B'}$ at right angles. By Theorem 13 the axis must also bisect $\overset{\frown}{AB}$ and $\overset{\frown}{A'B'}$. The axis of symmetry being a radius (Theos. 2, 4), we can state these results in the following way:

Theorem 16. *The radius which bisects an arc of a horocycle bisects the corresponding arc of any codirectional horocycle and also bisects both subtended chords at right angles.*

This theorem has several obvious implications worth noting. (a) The line joining the midpoints of two corresponding arcs or two corresponding chords is a radius. (b) The midpoints of two corresponding arcs and their subtended chords are collinear. (c) The line joining the midpoints of two corresponding chords is perpendicular to each chord.

Using the last-mentioned fact, one can easily show that $\overline{AA'}$ and $\overline{BB'}$ in Fig. V, 12 are equal and thus prove

Theorem 17. *Two codirectional horocycles intercept equal segments on their radii.*

By repeated use of the first part of Theorem 16 the following generalization of that part can be obtained.

Theorem 18. *Radii which divide an arc of a horocycle into n equal parts do likewise to the corresponding arc of any codirectional horocycle.*

In Fig. V, 12 the direction from A to A' on r is the direction of parallelism δ, and so is the direction from B to B' on s. For this reason we shall say that $\overline{A'B'}$ lies *in the direction of parallelism* from $\overline{AB}$ and that the direction from $\overline{A'B'}$ to $\overline{AB}$ is opposite to that of parallelism. A distinction between the two chords can thus be made. We shall distinguish between their subtended arcs in the same way.

Theorem 19. *Two corresponding chords of codirectional horocycles are always unequal, the lesser being the one which lies in the direction of parallelism from the other. The greater becomes arbitrarily great as it recedes in the direction opposite to that of parallelism.*

Proof. Let $\overline{AB}$, $\overline{A'B'}$ be the corresponding chords, $\overline{A'B'}$ lying in the direction of parallelism δ from $\overline{AB}$ (Fig. V, 13). If C, C' are the midpoints of $\overline{AB}$ and $\overline{A'B'}$, then line CC' meets these chords at right angles and is a radius (Theo. 16). Since lines AA', CC' are parallel in the direction δ and A'

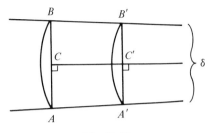

Fig. V, 13

lies in the direction of parallelism from A, the distance from A' to line CC' is less than the distance from A to this line (IV, §6, Theo. 20); that is, $A'C' < AC$. Hence $A'B' < AB$. Also, by the theorem just quoted, AC becomes arbitrarily great as A recedes in the direction opposite to δ. Hence AB likewise becomes arbitrarily great as A recedes in that direction.

Thus far it has not been necessary to deal with the *length* of an arc. Our ability to determine when two arcs of a horocycle were equal or unequal was based on the idea of congruent sets. But to proceed further in our work with such arcs we must deal with their length. The length of a horocyclic arc $\overset{\frown}{AB}$ is defined the same way as arc length is defined in Euclidean geometry. Let A_1, A_2, $\ldots$, A_{n-1} be $n - 1$ interior points of $\overset{\frown}{AB}$ such that A_1 is on $\overset{\frown}{AA_2}$, A_2 is on $\overset{\frown}{A_1A_3}$, $\ldots$, A_{n-1} is on $\overset{\frown}{A_{n-2}B}$ (Fig. V, 14). Consider the sum,

$$AA_1 + A_1A_2 + A_2A_3 + \cdots + A_{n-1}B,$$

of the lengths of the n chords $\overline{AA_1}$, $\overline{A_1A_2}$, $\ldots$, $\overline{A_{n-1}B}$. If, regardless of the

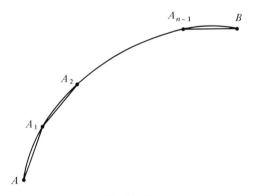

Fig. V, 14

particular points used, this sum has one and the same limit L when n becomes infinite and the length of each chord approaches zero, then L is called the *length of* $\overset{\frown}{AB}$. It can be shown that such a limit does exist and that horocyclic arc length has the same familiar properties as arc length in Euclidean geometry, but we shall not go into those arguments. Instead, we shall simply list the half-dozen or so properties that will be needed and accept them without proof,* that is, as assumptions.

Properties of Arc Length on a Horocycle

1. If a unit segment is chosen, then, in terms of this unit, every horocyclic arc $\overset{\frown}{AB}$ has a measure called its *length*. These measures are positive numbers. We shall denote the length of $\overset{\frown}{AB}$ by $s(\overset{\frown}{AB})$.

2. If a horocyclic arc is divided into a finite number of subarcs, the length of the arc equals the sum of the lengths of the subarcs. Hence the length of an arc always exceeds the length of a subarc.

3. If $\overset{\frown}{AB}$ is any horocyclic arc and λ is any positive number not exceeding $s(\overset{\frown}{AB})$, there is a unique point C on $\overset{\frown}{AB}$ such that $s(\overset{\frown}{AC}) = \lambda$.

4. If $\overset{\frown}{AB}$ (Fig. V, 15) is any horocyclic arc and $\overset{\frown}{A_n B_n}$, $n = 1, 2, 3, \ldots$, is a sequence of corresponding arcs such that A_n, B_n approach A, B when n becomes infinite, then $s(\overset{\frown}{A_n B_n})$ approaches $s(\overset{\frown}{AB})$ when n becomes infinite.

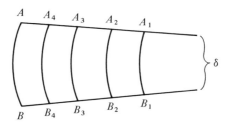

Fig. V, 15

5. If two arcs of the same or different horocycles are equal (that is, congruent), they have the same length, and vice versa. Hence unequal (that is, noncongruent) arcs have different lengths.

6. The length of a horocyclic arc exceeds the length of the subtended chord.

---

* The discoverers of hyperbolic geometry also accepted these properties without proof, but did not state them explicitly.

7.   The effect of changing from one unit segment to another x times as great is to multiply all horocyclic arc lengths by $1/x$. Thus the effect of halving the unit segment (that is, taking $x = \frac{1}{2}$) is to double all arc lengths.

The following theorem is just like Theorem 19 except that it deals with corresponding arcs. Its first sentence is an immediate consequence of Theorems 12 and 19. The second sentence involves the above-listed properties of arc length and its verification is left as an exercise.

Theorem 20.   *Two corresponding arcs of codirectional horocycles are always unequal, the lesser being the one which lies in the direction of parallelism from the other. The greater becomes arbitrarily great as it recedes in the direction opposite to that of parallelism.*

The remaining theorems of this section concern the ratios of arc lengths. Let $\overset{\frown}{AB}$ be an arc of a horocycle and let C be an interior point of $\overset{\frown}{AB}$. By the *ratio in which C (or the radius through C) divides* $\overset{\frown}{AB}$ we shall mean the number

$$\frac{s(\overset{\frown}{AC})}{s(\overset{\frown}{AB})}.$$

Thus the ratio in which the midpoint of the arc divides it is $\frac{1}{2}$ since $s(\overset{\frown}{AC})$ is half of $s(\overset{\frown}{AB})$. The two points which trisect the arc divide it in the ratios $\frac{1}{3}$ and $\frac{2}{3}$.

We now generalize the first part of Theorem 16.

Theorem 21.   *A radius which divides an arc of a horocycle will divide the corresponding arc of any codirectional horocycle so that the two ratios are equal.*

Proof.   Let $\overset{\frown}{AB}$, $\overset{\frown}{A'B'}$ be corresponding arcs of two horocycles with direction δ and let C be an inner point of $\overset{\frown}{AB}$ (Fig. V, 16). Since C is an inner point of $\overset{\frown}{AB}$, the radius $C\delta$ is between the radii $A\delta$ and $B\delta$, or, what is the same thing, between the radii $A'\delta$ and $B'\delta$. Combining this with the fact that radius $C\delta$ contains a point C' of the horocycle with arc $\overset{\frown}{A'B'}$, we conclude that C' is an interior point of $\overset{\frown}{A'B'}$. We must show that

$$\frac{s(\overset{\frown}{AC})}{s(\overset{\frown}{AB})} = \frac{s(\overset{\frown}{A'C'})}{s(A'B')}. \tag{1}$$

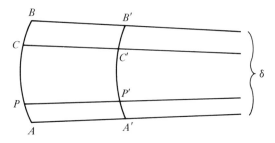

Fig. V, 16

Suppose, first, that $\overset{\frown}{AB}$ and $\overset{\frown}{AC}$ are commensurable (Fig. V, 16). This means that they are divisible into subarcs all of which are equal. If m of these subarcs constitute $\overset{\frown}{AC}$ and n of them constitute $\overset{\frown}{AB}$, then

$$s(\overset{\frown}{AC}) = m \cdot s(\overset{\frown}{AP}) \qquad \text{and} \qquad s(\overset{\frown}{AB}) = n \cdot s(\overset{\frown}{AP}), \qquad (2)$$

where $\overset{\frown}{AP}$ is the subarc with endpoint A. According to Theorem 18, the radii which divide $\overset{\frown}{AB}$ and $\overset{\frown}{AC}$ into these n and m equal subarcs also divide $\overset{\frown}{A'B'}$ and $\overset{\frown}{A'C'}$ into n and m equal subarcs. Hence

$$s(\overset{\frown}{A'C'}) = m \cdot s(\overset{\frown}{A'P'}) \qquad \text{and} \qquad s(\overset{\frown}{A'B'}) = n \cdot s(\overset{\frown}{A'P'}), \qquad (3)$$

where $\overset{\frown}{A'P'}$ corresponds to $\overset{\frown}{AP}$. From (2) and (3) we obtain

$$\frac{s(\overset{\frown}{AC})}{s(\overset{\frown}{AB})} = \frac{m}{n} \qquad \text{and} \qquad \frac{s(\overset{\frown}{A'C'})}{s(\overset{\frown}{A'B'})} = \frac{m}{n}.$$

Combining these equations gives (1).

Suppose, now, that $\overset{\frown}{AB}$ and $\overset{\frown}{AC}$ are not commensurable (Fig. V, 17). Let $\overset{\frown}{AC}$ be divided into m equal subarcs (see Ex. 3). Then $s(\overset{\frown}{AC}) = m \cdot s(\overset{\frown}{AP})$, where $\overset{\frown}{AP}$ is the subarc with endpoint A. Since $\overset{\frown}{AB}$, $\overset{\frown}{AC}$ are incommensurable, $\overset{\frown}{AB}$ is not divisible into arcs all of which equal $\overset{\frown}{AP}$, but is divisible into a certain whole number n of such equal arcs and one remaining shorter arc $\overset{\frown}{QB}$. That is, there is a point Q on $\overset{\frown}{AB}$ such that $s(\overset{\frown}{AQ}) = n \cdot s(\overset{\frown}{AP})$ and

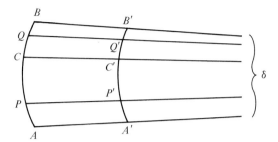

Fig. V, 17

$s(\overparen{QB}) < s(\overparen{AP})$. Since $\overparen{AQ}$ and $\overparen{AC}$ are commensurable, we have, by the first part of the proof,

$$\frac{s(\overparen{AC})}{s(\overparen{AQ})} = \frac{s(\overparen{A'C'})}{s(\overparen{A'Q'})}, \tag{4}$$

where Q' is the point in which the radius through Q meets $\overparen{A'B'}$. We note that $s(\overparen{Q'B'}) < s(\overparen{QB})$ by Theorem 20, and hence that $s(\overparen{Q'B'}) < s(\overparen{AP})$.

Now let P_1 be the midpoint of $\overparen{AP}$. Then $\overparen{AC}$ is divisible into $2m$ arcs each equal to $\overparen{AP_1}$. As before, $\overparen{AB}$ is not divisible into arcs all equal to $\overparen{AP_1}$, but is divisible into a whole number n_1 of such equal arcs and one remaining shorter arc $\overparen{Q_1B}$. That is, there is a point Q_1 on $\overparen{AB}$ such that $s(\overparen{AQ_1}) = n_1 \cdot s(\overparen{AP_1})$ and $s(\overparen{Q_1B}) < s(\overparen{AP_1})$. Arcs $\overparen{AC}$ and $\overparen{AQ_1}$ being commensurable, we obtain an equation like (4) except that Q_1, Q_1' replace Q, Q'. The whole procedure can be repeated by taking P_2 to be the midpoint of $\overparen{AP_1}$, then P_3 to be the midpoint of $\overparen{AP_2}$, and so forth. We thus obtain the sequence of equations

$$\frac{s(\overparen{AC})}{s(\overparen{AQ_i})} = \frac{s(\overparen{A'C'})}{s(\overparen{A'Q_i'})}, \qquad i = 1, 2, 3, \ldots, \tag{5}$$

where Q_i is on $\overparen{AB}$, Q_i' is the point in which the radius through Q_i meets $\overparen{A'B'}$, and $s(\overparen{Q_i'B'}) < s(\overparen{Q_iB}) < s(\overparen{AP_i})$.

Now let i become infinite. Then $s(\overparen{AP_i})$ approaches zero. According to
the inequalities at the end of the preceding paragraph, $s(\overparen{Q_i B})$ and $s(\overparen{Q_i' B'})$
must also approach zero. From the equation

$$s(\overparen{AQ_i}) = s(\overparen{AB}) - s(\overparen{Q_i B})$$

it then follows that $s(\overparen{AQ_i})$ approaches $s(\overparen{AB})$, and from

$$s(\overparen{A'Q_i'}) = s(\overparen{A'B'}) - s(\overparen{Q_i' B'})$$

that $s(\overparen{A'Q_i'})$ approaches $s(\overparen{A'B'})$. Consequently, the left side of (5) approaches
$s(\overparen{AC})/s(\overparen{AB})$ and the right side approaches $s(\overparen{A'C'})/s(\overparen{A'B'})$. These two ratios
must be equal since they are the limits of two sequences, represented by the
left and right sides of (5), all of whose terms are equal, respectively. Thus we
obtain (1). This completes the proof of Theorem 21.

The next theorem involves the ratio of the lengths of two corresponding
arcs of codirectional horocycles. To avoid a cumbersome statement of the
theorem we shall omit the words "of the length." Also, we shall understand
"ratio" to mean the ratio of the greater of the two arcs to the lesser. By the
distance between two corresponding arcs we shall mean the common length
of the radial segments intercepted by them (Theo. 17).

Theorem 22.   *The ratio of a pair of corresponding arcs of codirectional
horocycles depends only on the distance between them. More precisely, two
such ratios are equal if the two distances are equal, and they are unequal if the
two distances are unequal, the greater ratio corresponding to the greater
distance.*

Proof.   Let $\overparen{AB}$, $\overparen{A'B'}$ be corresponding arcs of two horocycles with
direction δ, and $\overparen{EF}$, $\overparen{E'F'}$ corresponding arcs of two horocycles with direction δ' (Fig. V, 18); δ and δ' need not be different. The distance between
$\overparen{AB}$, $\overparen{A'B'}$ is then AA' or BB', and the distance between $\overparen{EF}$, $\overparen{E'F'}$ is EE' or FF'.
We shall assume that $AA' = EE'$ and go on to show that

$$\frac{s(\overparen{AB})}{s(\overparen{A'B'})} = \frac{s(\overparen{EF})}{s(\overparen{E'F'})}. \tag{1}$$

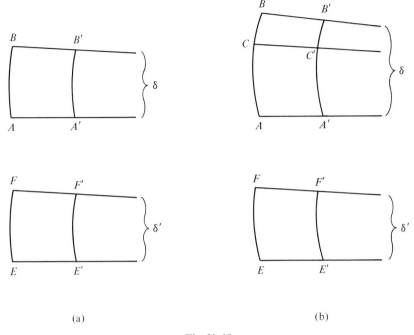

(a)   (b)

Fig. V, 18

If $\overset{\frown}{AB}$, $\overset{\frown}{EF}$ are equal (Fig. V, 18a), then $\overline{AB}$, $\overline{EF}$ are equal, so that tri-laterals $AB\delta$, $EF\delta'$ are congruent (Theo. 9). Hence quadrilaterals $ABB'A'$, $EFF'E'$ are congruent by side–angle–side–angle–side. Their remaining sides, $\overline{A'B'}$ and $\overline{E'F'}$, are therefore equal. Hence $\overset{\frown}{A'B'}$, $\overset{\frown}{E'F'}$ are equal. Equation (1) then follows.

If $\overset{\frown}{AB}$, $\overset{\frown}{EF}$ are unequal (Fig. V, 18b), let $\overset{\frown}{AB}$ be the greater. Take C on $\overset{\frown}{AB}$ so that $s(\overset{\frown}{AC}) = s(\overset{\frown}{EF})$. Let C' be the point in which the radius through C meets $\overset{\frown}{A'B'}$. By the first part of the proof

$$\frac{s(\overset{\frown}{AC})}{s(\overset{\frown}{A'C'})} = \frac{s(\overset{\frown}{EF})}{s(\overset{\frown}{E'F'})}. \tag{2}$$

By Theorem 21

$$\frac{s(\overset{\frown}{AC})}{s(\overset{\frown}{AB})} = \frac{s(\overset{\frown}{A'C'})}{s(\overset{\frown}{A'B'})}. \tag{3}$$

From (2) and (3) we obtain (1).

This completes the proof of the present theorem when $AA' = EE'$. The proof when $AA' \neq EE'$ is left to the student as Exercise 5.

Let $\overset{\frown}{AB}$, $\overset{\frown}{A'B'}$ be any two corresponding arcs of codirectional horocycles, $\overset{\frown}{A'B'}$ being the smaller. If we keep $\overset{\frown}{A'B'}$ fixed and let $\overset{\frown}{AB}$ recede in the direction opposite to that of the horocycles, then $s(\overset{\frown}{AB})$ becomes arbitrarily great (Theo. 20), and hence so does the ratio $s(\overset{\frown}{AB})/s(\overset{\frown}{A'B'})$. This implies that the ratio will exceed any number one may choose to mention. In particular, it will exceed e, the base of natural logarithms. We shall assume that there is a position of $\overset{\frown}{AB}$ such that the ratio actually has the value e. Use is made of this in the next section.

One further matter concerning these ratios is worth noting. If $\overset{\frown}{AB}$, $\overset{\frown}{A'B'}$ are two specific corresponding arcs, the values of $s(\overset{\frown}{AB})$ and $s(\overset{\frown}{A'B'})$ depend on the particular unit segment being used, but the value of the ratio $s(\overset{\frown}{AB})/s(\overset{\frown}{A'B'})$ does not—it is always the same regardless of the segment serving as unit. This is a consequence of arc length Property 7 stated earlier in this section. Thus, to give an example, if the ratio equals e for one unit segment, it will equal e for any other unit segment.

EXERCISES

1. Prove Theorem 17.

2. Using the first part of Theorem 16, prove Theorem 18.

3. Show that an arc of a horocycle can be divided into any number of equal parts. (Use the listed properties of arc length.)

4. Prove Theorem 20.

5. Complete the proof of Theorem 22 by considering the case in which $AA' \neq EE'$.

6. If $\overset{\frown}{AB}$, $\overset{\frown}{A'B'}$ are corresponding arcs of codirectional horocycles such that $s(\overset{\frown}{AB})/s(\overset{\frown}{A'B'}) = 5$, and $\overset{\frown}{MN}$ is the corresponding arc midway between them, find the value of $s(\overset{\frown}{AB})/s(\overset{\frown}{MN})$.

7. This is the same as Exercise 6 except that $\overset{\frown}{MN}$ is one-third of the way from $\overset{\frown}{AB}$ to $\overset{\frown}{A'B'}$.

8. If $\overset{\frown}{PQ}$, $\overset{\frown}{RS}$ are any two specific horocyclic arcs, show that the ratio $s(\overset{\frown}{PQ})/s(\overset{\frown}{RS})$ has the same value regardless of the unit segment.

6. ARC LENGTH ON A HOROCYCLE

So far we have considered only the *ratios* of arc lengths. Now we are in a position to obtain a formula for arc length itself.

We take any arc of a horocycle to serve as a reference, or initial, arc, denoting it by $\overparen{A_0 B_0}$ and its length by s_0. Then we consider any sequence of n smaller arcs $\overparen{A_1 B_1}$, $\overparen{A_2 B_2}$, ..., $\overparen{A_n B_n}$ corresponding to $\overparen{A_0 B_0}$ such that the distances $A_0 A_1, A_1 A_2, \ldots, A_{n-1} A_n$ are equal (Fig. V, 19). If $s_1, s_2, \ldots, s_n$

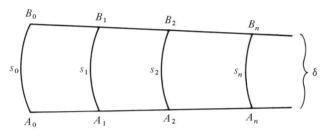

Fig. V, 19

denote the lengths of these n arcs, respectively, then, by Theorem 22, we have

$$\frac{s_0}{s_1} = \frac{s_1}{s_2} = \cdots = \frac{s_{n-1}}{s_n} = b, \tag{1}$$

where b is the number greater than 1 representing the common value of these ratios. Each of the lengths $s_1, s_2, \ldots, s_n$ can be expressed in terms of s_0 and b. Thus

$$
\begin{aligned}
s_1 &= s_0 b^{-1} \\
s_2 &= s_1 b^{-1} &= (s_0 b^{-1}) b^{-1} &= s_0 b^{-2} \\
s_3 &= s_2 b^{-1} &= (s_0 b^{-2}) b^{-1} &= s_0 b^{-3} \\
&\ \vdots &\vdots &\quad\vdots \\
s_n &= s_{n-1} b^{-1} &= (s_0 b^{-n+1}) b^{-1} &= s_0 b^{-n}
\end{aligned}
$$

or, in summary,

$$s_x = s_0 b^{-x}, \tag{2}$$

where $x = 1, 2, 3, \ldots, n$.

Before showing that this formula also holds for other values of x, related to the arcs situated between those we have considered, let us make two agreements which will render the formula more suitable for the important developments in the next section and the next chapter. First, let us agree to take for b the specific value e. This means that the arcs $\overparen{A_0 B_0}, \ldots, \overparen{A_n B_n}$ are such that the ratios in (1) are equal to the number e. This choice of arcs in no way depends on the unit segment, for, as we have already noted, the ratio of the lengths of any two specific arcs always has one and the same value regardless of the unit segment (also see §5, Ex. 8). Second, let us agree henceforth to take as unit segments the segments $\overline{A_0 A_1}, \overline{A_1 A_2}, \overline{A_2 A_3}, \ldots$ for which the ratios in (1) are equal to e. In other words, if the ratio of two corresponding arcs of codirectional horocycles is e, then each radial segment included by the arcs (or any segment congruent to it) will be our unit segment from now on. We shall call it the *standard unit segment* or the *standard unit of length*. When the formula $y = 2 \arctan(e^{-x})$, relating an angle of parallelism y to the corresponding distance x, was stated in Chapter IV, we mentioned that x was in terms of a certain unit segment but were in no position to describe the segment. It can now be said that the segment is the standard unit segment.

Because of the agreements just made formula (2) can be rewritten as

$$s_x = s_0 e^{-x}, \tag{3}$$

where x is any positive integer 1, 2, 3, $\ldots$, n, and s_x is the length of the arc $\overparen{A_x B_x}$ situated x units radially from $\overparen{A_0 B_0}$ in the direction of parallelism (Fig. V, 20). Retaining for s_x and $\overparen{A_x B_x}$ the meanings just stated, we shall

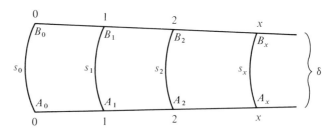

Fig. V, 20

now show that formula (3) holds not merely for positive integral values of x but for all positive values of x.

Consider, for example, the arc $\overparen{A_{1/2} B_{1/2}}$, which is $\frac{1}{2}$ unit from $\overparen{A_0 B_0}$ in the direction of parallelism, and whose length is denoted by $s_{1/2}$ (Fig. V, 21).

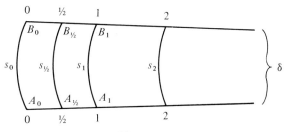

Fig. V, 21

Since $\overset{\frown}{A_{1/2}B_{1/2}}$ is also $\frac{1}{2}$ unit from $\overset{\frown}{A_1 B_1}$, we have, by Theorem 22,

$$\frac{s_0}{s_{1/2}} = \frac{s_{1/2}}{s_1}$$

$$(s_{1/2})^2 = s_0 s_1. \tag{4}$$

Since (3) holds when x is a positive integer, we know that $s_1 = s_0 e^{-1}$. Substituting this in (4) we get

$$(s_{1/2})^2 = s_0(s_0 e^{-1}) = s_0{}^2 e^{-1}$$

and hence

$$s_{1/2} = s_0 e^{-1/2}.$$

This proves that formula (3) holds for $x = \frac{1}{2}$.

We shall now prove that it holds for $x = \frac{1}{3}$ and $x = \frac{2}{3}$, using a method that can be readily applied to any positive fractional value of x. The arcs $\overset{\frown}{A_0 B_0}$, $\overset{\frown}{A_{1/3} B_{1/3}}$, $\overset{\frown}{A_{2/3} B_{2/3}}$, $\overset{\frown}{A_1 B_1}$ being equally spaced radially, we have

$$\frac{s_0}{s_{1/3}} = \frac{s_{1/3}}{s_{2/3}} = \frac{s_{2/3}}{s_1}$$

by Theorem 22. Denoting these ratios by c, we easily obtain

$$s_{1/3} = s_0 c^{-1}, \qquad s_{2/3} = s_0 c^{-2}, \qquad s_1 = s_0 c^{-3}. \tag{5}$$

This shows that s_0, $s_{1/3}$, $s_{2/3}$, s_1 form a geometric progression with first term s_0, last term s_1, and common ratio c^{-1}. To find c we substitute for s_1 in the last equation of (5), obtaining

$$s_0 e^{-1} = s_0 c^{-3}$$

$$c = e^{1/3}.$$

Putting this value into the first two equations of (5), we get

$$s_{1/3} = s_0 e^{-1/3} \qquad \text{and} \qquad s_{2/3} = s_0 e^{-2/3},$$

which proves formula (3) for $x = \frac{1}{3}$ and $x = \frac{2}{3}$. The proof of the formula for any positive fraction p/q is similar and is left as an exercise.

It remains to consider positive irrational values of x. Again, a simple example will suggest the general proof. Consider $x = \sqrt{2}$. Every irrational number is the limit of a sequence of rational numbers, in fact, of many such sequences. Of the sequences whose limit is $\sqrt{2}$, for example, the most familiar is 1, 1.4, 1.41, 1.414, ... If $\{x_n\}$ denotes any sequence of positive rational numbers with limit $\sqrt{2}$, then we can write $x_n \to \sqrt{2}$, meaning that x_n approaches $\sqrt{2}$ when n becomes infinite. When $x_n \to \sqrt{2}$, the points A_{x_n}, B_{x_n} approach the points $A_{\sqrt{2}}$, $B_{\sqrt{2}}$, respectively (Fig. V, 22). According

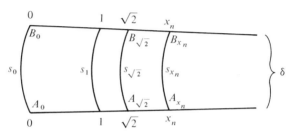

Fig. V, 22

to Property 4 of horocyclic arc length (§5), the length of $\overparen{A_{x_n} B_{x_n}}$ approaches that of $\overparen{A_{\sqrt{2}} B_{\sqrt{2}}}$ when $x_n \to \sqrt{2}$, or, in symbols,

$$\lim_{x_n \to \sqrt{2}} s_{x_n} = s_{\sqrt{2}}. \tag{6}$$

Since formula (3) holds for the positive rational number x_n, we have

$$s_{x_n} = s_0 e^{-x_n}. \tag{7}$$

Combining (6) and (7) gives

$$s_{\sqrt{2}} = \lim_{x_n \to \sqrt{2}} (s_0 e^{-x_n}). \tag{8}$$

The function $s_0 e^{-x}$ being continuous for all values of x, we have

$$\lim_{x_n \to \sqrt{2}} (s_0 e^{-x_n}) = s_0 e^{-\sqrt{2}}. \tag{9}$$

From (8) and (9) we get

$$s_{\sqrt{2}} = s_0 e^{-\sqrt{2}},$$

which shows that formula (3) holds for $x = \sqrt{2}$. Essentially the same argument can be used for any other positive irrational value of x. Thus we have

Theorem 23. *If s_0 is the length of any given arc of a horocycle, x is any positive number, and s_x is the length of the corresponding arc situated x units from the given arc in the direction of parallelism of the horocycle, then*

$$s_x = s_0 e^{-x}.$$

EXERCISES

1. Prove that formula (3) holds for the following values of x: (a) $\frac{3}{2}$; (b) $\frac{4}{3}$; (c) $\frac{1}{4}$; (d) p/q, where p, q are positive integers; (e) $\sqrt{3}$.

2. By considering the arcs which correspond to $\overparen{A_0 B_0}$ (Fig. V, 21) and lie to the left of it, that is, in the direction opposite to δ, one can show that formula (3) also holds for negative values of x. Verify this for the following values of x: (a) -1, (b) -2, (c) $-\frac{1}{2}$.

3. If the length of the given arc in Theorem 23 is 1, find the length of the corresponding arc 1 unit away in the direction of parallelism of the horocycle.

7. FORMULAS RELATED TO k-ARCS

Using Theorem 23 we shall now obtain four important formulas involving the k-arcs considered in Section 4. These arcs, which were defined to be certain horocyclic arcs of the same length, were shown to have the property that the tangent at either endpoint is parallel to the radius through the other endpoint in the direction along the radius which is opposite to that of the horocycle (Theo. 15). Thus, if $\overparen{AD}$ is any k-arc of a horocycle with direction δ (Fig. V, 23), then the tangent at A is parallel, in the direction δ', to the radius through D. (The tangent is line AC.) The length of all k-arcs will be denoted by S.

Proceeding to the derivation of the four formulas, we let B denote any interior point of the k-arc $\overparen{AD}$ (Fig. V, 23). The radius through B will meet line $A\delta'$, the tangent at A, in a point C. (This is seen by noting that the radius must meet chord $\overline{AD}$, which is the middle side of trilateral $AD\delta'$, and by applying IV, §4, Theo. 11.) We denote the three sides of the curvilinear figure ABC by r, s, t, where r is the radial distance BC, s is the length of $\overparen{AB}$, and t is the tangential distance AC. The formulas we seek are equations relating r, s, t, and S. Since $\overparen{AB}$ is a subarc of $\overparen{AD}$, we already know that $s < S$. Also, by considering chord $\overline{AB}$ and noting that $\measuredangle AB\delta$ is acute, we infer that $\measuredangle ABC$ in triangle ABC is obtuse and hence that $r < t$. In Fig. V, 23 we have put arrows on straight lines to help distinguish them from arcs.

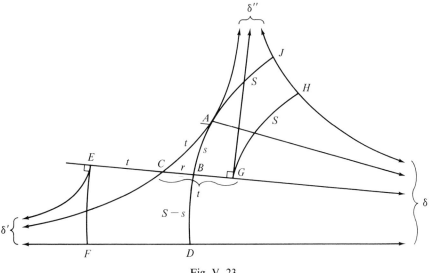

Fig. V, 23

Now consider trilateral $AC\delta$. Since $\measuredangle CA\delta$ is a right angle, $\measuredangle AC\delta$ must be the angle of parallelism for the distance t, and, of course, so is the angle which is vertical to $\measuredangle AC\delta$. Hence if we take E on line BC to the left of C so that $CE = t$, the perpendicular to line BC at E is parallel to line AC in the direction δ' (that is, the perpendicular is line $E\delta'$). The horocycle (E, δ) has a point F on the radius $D\delta$, and the tangent to this horocycle at E is line $E\delta'$, which is parallel to the radius $D\delta$ in the direction δ'. Hence $\overset{\frown}{EF}$ is a k-arc and its length is S. Applying Theorem 23 to the corresponding arcs $\overset{\frown}{BD}$ and $\overset{\frown}{EF}$, we obtain

$$S - s = Se^{-(t+r)}. \tag{1}$$

This equation resulted from working with the point E on line BC which is t units from C and to the left of it. Let us now work with the point on that line which is t units from C and to the right of it, namely, G. For the reason given at the beginning of the preceding paragraph, the perpendicular to line BC at G is parallel to line AC in the direction δ''. This perpendicular, line $G\delta''$, is tangent at G to the horocycle (G, δ). If we take H on this horocycle so that $\overset{\frown}{GH}$ is a k-arc, then the radius through H is parallel to line $G\delta''$ in the direction δ''. The horocycle (A, δ), of which $\overset{\frown}{AB}$ is a part, contains a point J on that radius. Since the tangent to the horocycle (A, δ) at A is parallel in the direction δ'' to the radius through J, $\overset{\frown}{AJ}$ is a k-arc. It follows that $\overparen{GH}$ and $\overparen{BJ}$ are corresponding arcs with lengths S and $S + s$, respectively, and at a distance $t - r$ from each other. Hence, by Theorem 23,

$$S = (S + s)e^{-(t-r)},$$

or

$$S + s = Se^{t-r}. \tag{2}$$

Equations (1) and (2) can be expressed more concisely. Adding them and dividing the result by S gives

$$2 = e^{-t-r} + e^{t-r},$$

which reduces to

$$e^r = \frac{(e^t + e^{-t})}{2},$$

or

$$e^r = \cosh t \tag{3}$$

by definition of cosh t (called the *hyperbolic cosine of t*).
     Subtracting (1) from (2) and dividing the result by 2 gives

$$s = Se^{-r}\frac{(e^t - e^{-t})}{2},$$

or

$$s = Se^{-r}\sinh t$$

by definition of sinh t (called the *hyperbolic sine of t*). Substituting from (3) into the last equation we then obtain

$$s = S\frac{\sinh t}{\cosh t},$$

or

$$s = S \tanh t \tag{4}$$

by definition of tanh t (called the *hyperbolic tangent of t*).
     Equations (3) and (4) are two of the four formulas we planned to derive. Before proceeding to the others, which require a different diagram, let us state what has been proved.

Theorem 24. *If $\overset{\frown}{AB}$ is a horocyclic arc shorter than a k-arc and C is the point in which the tangent at A meets the radius through B* (Fig. V, 24), *then*

$$e^r = \cosh t \qquad \text{and} \qquad s = S \tanh t,$$

where s is the length of $\overset{\frown}{AB}$, r is the radial distance BC, t is the tangential distance AC, and S is the length of a k-arc.

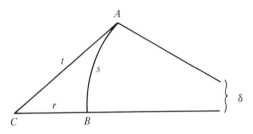

Fig. V, 24

Now consider *any* arc $\overset{\frown}{OP}$ of a horocycle with direction δ (Fig. V, 25). Let Q be the projection of P on the radius through O. Our concern is with the three-sided figure OPQ. Denote the length of $\overset{\frown}{OP}$ by s, that of $\overline{OQ}$ by x, and that of $\overline{PQ}$ by y. The two formulas to be derived involve x, y, and s. The codirectional horocycle (Q, δ) has a point R on the radius $P\delta$. Since $\overset{\frown}{OP}$, $\overset{\frown}{QR}$ are corresponding arcs, the length of $\overline{PR}$ is x. Line PQ, being perpendicular to the radius $Q\delta$, is tangent to $\overset{\frown}{QR}$ at Q. Thus the tangent to $\overset{\frown}{QR}$ at Q meets the radius through R at P. Therefore $\overset{\frown}{QR}$ is not a k-arc.

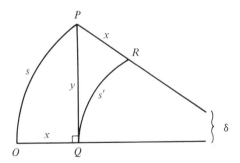

Fig. V, 25

Also, it is not longer than a *k*-arc, for if it were, the tangent at Q would not meet the radius through R (this is not difficult to show and is left as Ex. 1). Hence $\overparen{QR}$ is shorter than a *k*-arc. Theorem 24 therefore applies to the figure PQR and we have

$$e^x = \cosh y. \tag{5}$$

This equation, of course, also relates parts of the figure OPQ. It is therefore one of the two formulas we are seeking.

To obtain the other, let s' be the length of $\overparen{QR}$. Then, by Theorem 24,

$$s' = S \tanh y.$$

Since $\overparen{OP}$, $\overparen{QR}$ are corresponding arcs, we have, by Theorem 23,

$$s' = se^{-x}.$$

Combining the last two equations gives

$$se^{-x} = S \tanh y$$
$$s = S(\tanh y)e^x,$$

which, after substitution from (5), becomes

$$s = S \tanh y \cosh y,$$

or

$$s = S \sinh y. \tag{6}$$

This is the desired formula.

Theorem 25. *If $\overparen{OP}$ is any arc of a horocycle and Q is the projection of P on the radius through O (Fig. V, 25),* then

$$e^x = \cosh y \qquad \text{and} \qquad s = S \sinh y,$$

*where s is the length of $\overparen{OP}$, x the distance OQ, y the distance PQ, and S the length of a *k*-arc.*

Computations with the formulas of Theorems 24 and 25 usually require the use of tables giving the values of the exponential and hyperbolic functions involved. For example, to find y from the formula $e^x = \cosh y$ when $x = 2$, we have $\cosh y = e^2 = 7.4$ (to one decimal place by tables), so that $y = 2.6$ (to one decimal place by tables).

Looking back at equation (2) of the preceding section, we can now see that the two agreements made at that time, namely, to take for b the specific value e and to use the standard unit segment, enabled us to bring the hyperbolic functions into the discussion of the present section and thus achieve formulas of great simplicity in Theorems 24 and 25. These formulas play an important part in the work of the next chapter.

EXERCISES

1. If $\overset{\frown}{AB}$ is a horocyclic arc longer than a k-arc, show that the tangent at A does not meet the radius through B.

2. In Theorem 24, (a) calculate r when $t = 1$; (b) calculate t when $r = 2$.

3. In Theorem 24, (a) calculate t when $s = \frac{1}{2}S$; (b) how does t vary when s increases and approaches S?

4. Calculate x and y in Theorem 25 if $\overset{\frown}{OP}$ is a k-arc.

5. Show (by reference to tables or by mathematical proof) that $0 < \tanh t < 1$ for all positive values of t, and hence that the formula $s = S \tanh t$ in Theorem 24 is consistent with the hypothesis that $\overset{\frown}{AB}$ is shorter than a k-arc.

6. It can be proved that $\sinh y$ takes on all positive values when $y > 0$. Using this fact, show that the formula $s = S \sinh y$ in Theorem 25 is consistent with the hypothesis that $\overset{\frown}{OP}$ is *any* horocyclic arc.

7. From an endpoint A of a horocyclic arc a perpendicular is dropped to the radius through the other endpoint B. If $\overset{\frown}{AB}$ is half of a k-arc, find the length of the perpendicular and the distance from B to the foot of the perpendicular.

Triangle Relations

1. INTRODUCTION

Our main objective in this chapter is to obtain the formulas of hyperbolic geometry which express the numerical relations among the sides and angles of a triangle. Each of these formulas is analogous to some familiar Euclidean formula. Corresponding to the right triangle formulas $a^2 + b^2 = c^2$ and $\sin A = a/c$, for example, there are two hyperbolic formulas also involving the quantities a, b, c and a, c, A, respectively, but relating them in a different way. To derive the new formulas, we shall need, in addition to the facts about horocycles learned in the preceding chapter, some knowledge of what are called *associated right triangles*. Therefore we discuss them first.

2. ASSOCIATED RIGHT TRIANGLES

Let ABC (Fig. VI, 1) be any right triangle, labeled in the usual way; that is, the right angle is at C and the sides opposite A, B, C have the lengths a, b, c. Let the angles at A and B have the measures λ and μ. We shall prove that the existence of this triangle implies the existence of four other right triangles. The five triangles are therefore said to be *associated* with each other. The sides and angles of the four triangles have measures which are related to a, b, c, λ, μ and in order to obtain suitable symbols for them we make the following agreements:

1. α, β, γ denote the angles of parallelism* corresponding to a, b, c and

---

* In this chapter when we say "angle of parallelism" we shall mean the measure of the angle.

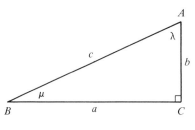

Fig. VI, 1

l, m denote the distances corresponding to λ, μ, regarded as angles of parallelism.

2. α', β', γ', λ', μ' denote the complements of α, β, γ, λ, μ.

3. a', b', c', l', m' denote the distances corresponding to α', β', γ', λ', μ', regarded as angles of parallelism.

4. The distances a', b', c', l', m' will be called *complementary* to the distances a, b, c, l, m, and vice versa, since the corresponding angles of parallelism are complementary. For example, a and a' are called complementary because $\alpha + \alpha' = 90°$.

Agreements 1, 2, and 3 are summarized in the following table.

Distance:	a	b	c	l	m	a'	b'	c'	l'	m'
Angle of parallelism:	α	β	γ	λ	μ	α'	β'	γ'	λ'	μ'

Since triangle ABC is given, the values a, b, c, λ, μ are presumed known. From them the values α, β, γ, l, m are then determined by the formula

$$y = 2 \arctan(e^{-x}), \tag{1}$$

where y is the angle of parallelism corresponding to the distance x. Thus $\alpha = 2 \arctan(e^{-a})$ and similar equations give the values of β and γ. If formula (1) is solved for x, giving

$$x = \log_e \cot \frac{y}{2}, \tag{2}$$

then $l = \log_e \cot(\lambda/2)$ and a similar equation provides the value of m. The values α, β, γ, l, m can therefore be regarded as known. From them and λ, μ the values α', β', γ', λ', μ' can be found. Finally, from these values, by use of (2), the values a', b', c', l', m' can be found. Thus, given triangle ABC, the 20 values listed in the table above may be thought of as known and we shall therefore feel free to use them in obtaining the other associated triangles.

Regarding the given triangle as associated Triangle 1, let us proceed to obtain Triangle 2. This is done with the aid of a Lambert quadrilateral in which the two sides adjacent to a right angle have lengths c and m'. (It is easily shown that such quadrilaterals exist. This is left as Ex. 1.) Let $DEFG$ (Fig. VI, 2) be such a quadrilateral, with $DE = c$, $EF = m'$, an acute angle at D, and right angles at E, F, G. We shall need to find the measures of the other parts of the quadrilateral. Let H be the point on $\overline{DG}$ such that $FH = c$. Then, according to the procedure for constructing boundary parallels (IV, §3, Ex. 4), line FH is the right-hand parallel to line DE through F and $\measuredangle EFH$ is the angle of parallelism for the distance EF. Thus $\measuredangle EFH = \mu'$. Hence $\measuredangle GFH = \mu$. It follows that triangle HFG is congruent to triangle ABC.

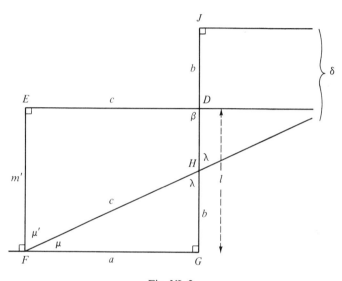

Fig. VI, 2

Therefore $FG = a$, $GH = b$, and $\measuredangle FHG = \lambda$. Also, $DG = l$ [this will be seen to follow from IV, §3, Theorem 9, part (b), on comparing Fig. VI, 2 with Fig. IV, 15]. It remains to find the value of $\measuredangle EDG$. Extend $\overline{DG}$ to J so that $DJ = b$. Since $HJ = l$, the perpendicular to $\overline{HJ}$ at J is parallel to line FH in the direction δ. It is therefore also parallel to line DE in that direction. From this we conclude that $\measuredangle EDG = \beta$.

The measures of all the sides and the acute angle of the quadrilateral are now known and, as the following table helps to make clear, each is related to some measure of the given triangle. The quadrilateral is labeled Quadrilateral 1 and the arrow reminds us that it has been obtained from Triangle 1:

$$\begin{matrix} \text{Triangle 1:} & a, & b, & c, & \lambda, & \mu \\ \text{Quadrilateral 1:} & a, & \beta, & c, & l, & m'. \end{matrix} \qquad (3)$$

We now exhibit Quadrilateral 1 as in Fig. VI, 3, so that the sides labeled
l, *a* occupy the positions previously occupied by the sides labeled *c*, *m'*, and
shall refer to it thus exhibited as Quadrilateral 1, Reversed. Proceeding with
this quadrilateral just as we did with its counterpart in Fig. VI, 2, we obtain
a right triangle *FEK* analogous to triangle *FGH*. Triangle *FEK* is our
associated Triangle 2. Its measures *m'*, *b*, *l*, *γ*, *α'* have the same relation to

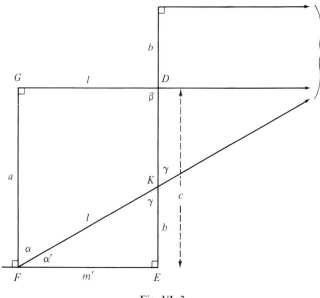

Fig. VI, 3

those of Quadrilateral 1, Reversed, as do the measures of triangle *FGH*
(or Triangle 1) to those of Quadrilateral 1. The following table helps to make
this clear:

$$
\begin{array}{llllll}
\text{Quadrilateral 1, Reversed:} & m', & \beta, & l, & c, & a \\
\text{Triangle 2:} & m', & b, & l, & \gamma, & \alpha'.
\end{array}
\qquad (4)
$$

In (4), as in (3), the first two measures represent the arms of the triangle, the
last two represent the angles opposite them, respectively, and the third
represents the hypotenuse.

To obtain another associated triangle by the method being used clearly
requires a new quadrilateral. Let us return to Triangle 1 for a moment and
note that if we had originally listed its parts in the order

$$
b, \quad a, \quad c, \quad \mu, \quad \lambda,
$$

we would have obtained from this Triangle 1, Reversed, a quadrilateral with parts

$$b, \quad \alpha, \quad c, \quad m, \quad l',$$

and hence different from Quadrilateral 1. This new quadrilateral could be used to obtain a third associated triangle, but it is preferable to use for that purpose the new quadrilateral which can be deduced from Triangle 2, Reversed. That quadrilateral is labeled Quadrilateral 2 in the following table:

$$\text{Triangle 2, Reversed:} \quad b, \quad m', \quad l, \quad \alpha', \quad \gamma$$
$$\text{Quadrilateral 2:} \quad b, \quad \mu', \quad l, \quad a', \quad c'.$$

Just as Triangle 2 was obtained from Quadrilateral 1, Reversed, so now we obtain Triangle 3, our third associated triangle, from Quadrilateral 2, Reversed:

$$\text{Quadrilateral 2, Reversed:} \quad c', \quad \mu', \quad a', \quad l, \quad b \tag{5}$$
$$\text{Triangle 3:} \quad c', \quad m', \quad a', \quad \lambda, \quad \beta'.$$

By repeating the procedure described above, that is, by reversing Triangle 3, and so on, we would obtain

$$\text{Triangle 4:} \quad l', \quad c', \quad b', \quad \alpha', \quad \mu \tag{6}$$

as our next associated triangle, and then

$$\text{Triangle 5:} \quad a, \quad l', \quad m, \quad \beta', \quad \gamma. \tag{7}$$

As before, the first two measures in each of the listings (6) and (7) represent the arms, the last two represent the angles opposite them, respectively, and the third represents the hypotenuse.

There is nothing to prevent a continuation of the procedure beyond Triangle 5. It turns out, however, that the sixth triangle is congruent to Triangle 1, with the result that the seventh is necessarily congruent to Triangle 2, the eighth to Triangle 3, and so on. Hence, essentially no more than five triangles can be obtained by the method we have used, namely, Triangles 1 through 5 above. Examination of their measures shows that these triangles are generally distinct. Our result can therefore be stated as follows.

Theorem 1. *Let any right triangle be given, with sides and acute angles having the measures listed in (3). The existence of this triangle implies the existence of the four right triangles whose corresponding measures are listed in (4) through (7). The five triangles are generally distinct. (They are called associated triangles.)*

There is a simple device that enables us to write down the measures of the five triangles without having to memorize them all. Take five evenly spaced points on a circle and label them c, l, a', b', m as in Fig. VI, 4. These are the measures of the hypotenuses of the triangles. Choose any one of these measures, say c. Because of their position on the circle relative to c, we shall call l, m *adjacent measures* and a', b' *opposite measures*. Then we

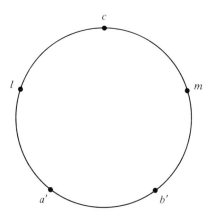

Fig. VI, 4

note that the complements of these opposite measures are a, b and that the angles of parallelism corresponding to the adjacent measures are λ, μ. Thus we obtain the measures of the arms and the acute angles of the triangle with hypotenuse c (Triangle 1). Similarly, to write down the measures of any other associated triangle, we choose the letter on the circle that represents the hypotenuse of that triangle. Then the complements of the opposite measures will represent the arms of the triangle and the angles of parallelism corresponding to the adjacent measures will represent the acute angles of the triangle. For example, if m is chosen, then the opposite measures are a', l and the adjacent measures are b', c, so that the arms of the triangle with hypotenuse m have the values a, l' and the acute angles have the values β', γ.

EXERCISES

1. Calculate the following:

(a) α', if $\alpha = 60°$;   (b) γ, if $\gamma' = 20°$;   (c) β, if $b = 1$;
(d) l, if $\lambda = \pi/3$;   (e) μ', if $m' = 3$;   (f) m, if $m' = 3$.

2. Show how to obtain a Lambert quadrilateral in which the two sides adjacent to a right angle have the given lengths c and m'.

3. Prove that all Lambert quadrilaterals which meet the condition specified in Exercise 2 are congruent.

4. Show that associated Triangles 1 and 2 are congruent if λ equals the angle of parallelism for the distance c, and only then.

5. Are associated Triangles 2 and 3 ever congruent? Justify your answer.

6. Show in detail how to obtain (a) Triangle 4 from Triangle 3; (b) Triangle 5 from Triangle 4.

7. Obtain the sixth triangle from Triangle 5 and show that it is congruent to Triangle 1.

8. Actually, the existence of any one of the five associated triangles implies the existence of the others. Why is this so?

3. IMPROVED ANGLE OF PARALLELISM FORMULAS

We shall now change the equation

$$y = 2\arctan(e^{-x}), \tag{1}$$

where y is the angle of parallelism corresponding to the distance x, to a form which is more concise, simplifies the calculation of y, and is also helpful in other ways.

Let us make the substitution

$$z = \arctan(e^{-x}) \tag{2}$$

in (1). We can then write

$$y = 2z$$

$$\tan y = \tan 2z = \frac{2\tan z}{1 - \tan^2 z} \tag{3}$$

by a standard trigonometric formula. From (2) we have

$$\tan z = e^{-x}.$$

Substituting this in (3) gives

$$\tan y = \frac{2e^{-x}}{1 - e^{-2x}} = \frac{2}{e^x - e^{-x}} = \frac{1}{\sinh x} = \operatorname{csch} x,$$

by definition of $\operatorname{csch} x$ (called the *hyperbolic cosecant of* x) as the reciprocal of $\sinh x$.

The equation

$$\tan y = \operatorname{csch} x \tag{4}$$

is the simplified form of (1) that we wished to obtain. From it we can deduce other equally simple and useful forms. First, taking reciprocals, we obtain

$$\cot y = \sinh x. \tag{5}$$

From this and the fact that $1 + \sinh^2 x = \cosh^2 x$ (proof of this is left as Ex. 1), we can then write

$$\csc y = (1 + \cot^2 y)^{1/2} = (1 + \sinh^2 x)^{1/2} = \cosh x. \tag{6}$$

Next, from (6) we obtain

$$\sin y = \frac{1}{\csc y} = \frac{1}{\cosh x} = \operatorname{sech} x, \tag{7}$$

by definition of $\operatorname{sech} x$ (called the *hyperbolic secant of* x) as the reciprocal of $\cosh x$.

Using (7) and the fact that $1 - \operatorname{sech}^2 x = \tanh^2 x$ (proof of this is left as Ex. 2), we can write

$$\cos y = (1 - \sin^2 y)^{1/2} = (1 - \operatorname{sech}^2 x)^{1/2} = \tanh x. \tag{8}$$

Finally, taking reciprocals in (8) gives

$$\sec y = \frac{1}{\tanh x} = \coth x, \tag{9}$$

by definition of $\coth x$ (called the *hyperbolic cotangent of* x) as the reciprocal of $\tanh x$.

We summarize as follows:

Theorem 2. *If y is the angle of parallelism corresponding to the distance x, then*

$$\begin{array}{ll} \sin y = \operatorname{sech} x & \csc y = \cosh x \\ \cos y = \tanh x & \sec y = \coth x \\ \tan y = \operatorname{csch} x & \cot y = \sinh x. \end{array}$$

These equations can be used immediately to help find a formula relating two complementary distances. Thus far we have no such formula. Given the distance b, for example, we can now find the complementary distance b' only by first determining the angle of parallelism β corresponding to b, then finding the complementary angle β', and finally calculating the distance b' for which β' is the angle of parallelism. We shall obtain the desired formula simply by using general symbols and combining the equations resulting from the preceding steps.

Let x be any distance and y the corresponding angle of parallelism. Then, from Theorem 2,

$$\sinh x = \cot y. \tag{10}$$

If y' is the angle complementary to y, we know that

$$\cot y = \tan y'. \tag{11}$$

Let x' be the distance corresponding to y' as angle of parallelism. Then, by Theorem 2,

$$\tan y' = \operatorname{csch} x'. \tag{12}$$

On combining (10), (11), and (12), we obtain

$$\sinh x = \operatorname{csch} x', \tag{13}$$

which is the desired formula relating the complementary distances x and x'. Two other equally simple and useful formulas relating x and x' can be deduced from (13). We state them in the following theorem and leave their derivation as an exercise.

Theorem 3. *If x and x' are two complementary distances, then*

$$\sinh x = \operatorname{csch} x'$$
$$\cosh x = \coth x'$$
$$\tanh x = \operatorname{sech} x'.$$

EXERCISES

1. Prove the formula $1 + \sinh^2 x = \cosh^2 x$.

2. From the formula of Exercise 1 deduce the formulas $1 - \operatorname{sech}^2 x = \tanh^2 x$ and $1 + \operatorname{csch}^2 x = \coth^2 x$.

3. Derive the last two equations in Theorem 3.

4. Using Theorem 2 calculate (a) the angle of parallelism corresponding to a distance of 1; (b) the distance corresponding to an angle of parallelism of 30°.

5. Use Theorem 3 to calculate the distance which is complementary to a distance of 1. Verify that it is immaterial whether we substitute 1 for x or for x'.

4. REMARKS ON THE TRIGONOMETRIC FUNCTIONS

As we have seen, all the equations relating angles of parallelism and their corresponding distances involve trigonometric functions or their inverses. Our derivations of these equations in the preceding section depended on familiar trigonometric formulas, and the student's calculations with the equations in exercises involved the use of familiar trigonometric tables. It would appear, then, that the facts concerning the trigonometric functions are the same in hyperbolic geometry as in Euclidean geometry. This is correct, but the matter calls for further discussion if the student is not to draw wrong conclusions from this statement. He may think, for example, that in hyperbolic geometry, as in Euclidean geometry, the sine of an acute angle is the ratio of the side opposite the angle to the hypotenuse in a right triangle containing the angle. This is not true, as we prove later.

Historically, trigonometry developed within the framework of Euclidean geometry and was made possible by the existence of triangles that are similar but not congruent. Thus, if $\angle A$ is acute, $\sin A$ was defined originally as the ratio

$$\frac{\text{length of side opposite } \angle A}{\text{length of hypotenuse}}$$

in any right triangle containing $\angle A$, and corresponding familiar definitions were made for $\cos A$, $\tan A$, and the other trigonometric functions. By use of these definitions trigonometric tables were constructed and basic formulas such as $\tan A = \sin A/\cos A$, $\sin^2 A + \cos^2 A = 1$, $\sin 2A = 2 \sin A \cos A$, and so on, were proved. Later, the definitions were broadened so as to apply to nonacute angles. This, too, was done within a Euclidean framework, and in such a way that the tables and basic formulas continued to hold.

The study of the trigonometric functions remained a part of numerical Euclidean geometry down to modern times. Then it was discovered that these functions could be represented by infinite series. Thus, if x is any number of standard angular units, then

$$\sin x = x - \frac{x^3}{3!} + \frac{x^5}{5!} - \frac{x^7}{7!} + \cdots \tag{1}$$

$$\cos x = 1 - \frac{x^2}{2!} + \frac{x^4}{4!} - \frac{x^6}{6!} + \cdots \tag{2}$$

$$\tan x = x + \frac{x^3}{3} + \frac{2x^5}{15} + \cdots, \tag{3}$$

and there are corresponding series for csc x, sec x, cot x. This discovery was important, for it offered the possibility of calculating trigonometric values without directly using triangles, and hence of obtaining greater accuracy in those values. (Trigonometric tables are now actually obtained by calculations based on such series.) Moreover, it was also discovered that all the basic trigonometric formulas could be deduced from these series by purely numerical methods. In other words, it became clear that the entire theory of the trigonometric functions could be developed independently of Euclidean geometry if those functions were redefined by means of infinite series. When that is done, the resulting branch of mathematics is known as *trigonometric analysis.*

This subject is concerned primarily with the numerical aspects of the trigonometric functions and its elementary facts are therefore familiar to the student. An example of a fact belonging to trigonometric analysis is the statement that $\sin(\pi/6) = \frac{1}{2}$, where $\pi/6$ is simply a number (of approximate value 0.52). When this fact is applied in geometry, Euclidean or hyperbolic, we interpret $\pi/6$ as a number of standard angular units, but still say $\sin(\pi/6) = \frac{1}{2}$ (or, what is equivalent, $\sin 30° = \frac{1}{2}$). All trigonometric values are to be regarded in the same way. It is not a fact of trigonometric analysis, however, that in a Euclidean right triangle containing an angle of $\pi/6$ the ratio of the opposite side to the hypotenuse has the value $\sin(\pi/6)$, or $\frac{1}{2}$. This is a fact of Euclidean geometry. In the preceding section we derived formulas relating angles of parallelism and their corresponding distances. These formulas are facts of hyperbolic geometry. But the meanings and values of $\sin y$, $\cos y$, $\tan y$, and so on, which appear in these formulas, and the identities $\sin y = 1/\csc y$, $\sin^2 y + \cos^2 y = 1$, and so on, used in their derivation are facts of trigonometric analysis and hence entirely independent of hyperbolic geometry.

The same remarks apply to the formulas relating the sides and angles of a triangle which will be derived in the sections to follow. In the case of a right triangle, for example, one of these formulas is

$$\sin \lambda = \frac{\sinh a}{\sinh c}.$$

This is a fact of hyperbolic geometry and it is analogous to the formula

$$\sin \lambda = \frac{a}{c},$$

which is a fact of Euclidean geometry. In both formulas the meaning of the left side is not a fact of either system of geometry, but of trigonometric analysis.

5. RIGHT TRIANGLE FORMULAS

Consider a right triangle labeled A, B, C, a, b, c, λ, μ (Fig. VI, 5) as in Section 2. The distance l, which corresponds to λ regarded as an angle of parallelism, may exceed c, be less than c, or be equal to c. We shall consider only the case $l > c$, leaving the others to be handled in exercises.

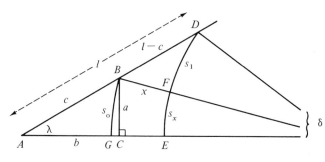

Fig. VI, 5

On $\overline{AB}$ extended take point D so that $AD = l$. The line perpendicular to $\overline{AD}$ at D is then parallel to line AC in the direction δ. Let E, F be the points in which the horocycle (D, δ) meets its radii $A\delta$, $B\delta$, and let G be the point in which the codirectional horocycle (B, δ) meets its radius $A\delta$. Line AD, we note, is tangent to the horocycle (D, δ) at D.

Denote the length of $\overarc{BG}$ by s_0, the length of $\overarc{EF}$ by s_x, and the distance BF by x. Then, from Theorems 23 through 25 of Chapter V we obtain, respectively,

$$s_x = s_0 e^{-x} \tag{1}$$

$$e^x = \cosh(l - c) \tag{2}$$

$$s_0 = S \sinh a. \tag{3}$$

Also, if we denote the length of $\overarc{DF}$ by s_1, we obtain from Theorem 24

$$s_1 = S \tanh(l - c) \tag{4}$$

$$s_1 + s_x = S \tanh l. \tag{5}$$

Combining equations (1) through (5), starting with (3), gives

$$\sinh a = \frac{s_0}{S} = \frac{s_x e^x}{S} = e^x \left(\frac{s_1 + s_x}{S} - \frac{s_1}{S} \right)$$

$$= \cosh(l - c)[\tanh l - \tanh(l - c)]$$

$$= \tanh l \cosh(l - c) - \sinh(l - c)$$

$$= \frac{\sinh l \cosh(l - c) - \cosh l \sinh(l - c)}{\cosh l}$$

$$= \frac{\sinh c}{\cosh l},$$

the last numerator being the result of using the formula

$$\sinh(A - B) = \sinh A \cosh B - \cosh A \sinh B,$$

where A, B are any numbers. (The proof of this formula is covered in an exercise.) Since $\cosh l = \csc \lambda$ by Theorem 2 in Section 3, we obtain, finally,

$$\sinh a = \frac{\sinh c}{\csc \lambda},$$

or

$$\sin \lambda = \frac{\sinh a}{\sinh c}. \tag{6}$$

This formula expresses the relation between the acute angle λ, the opposite side a, and the hypotenuse c,* and is therefore analogous to the Euclidean formula $\sin \lambda = a/c$, as was mentioned earlier. Since ABC is any right triangle and λ is really any one of its acute angles, our result can be stated as follows:

Theorem 4.   *The sine of an acute angle in a right triangle is equal to the hyperbolic sine of the opposite side divided by the hyperbolic sine of the hypotenuse.*

Applying this theorem to the acute angle μ in triangle ABC, which is the angle opposite side b, we obtain

$$\sin \mu = \frac{\sinh b}{\sinh c}. \tag{7}$$

---

* Of course, we mean that a and c are the *measures* of the opposite side and hypotenuse. For conciseness of expression we shall continue to omit the word "measure" in statements like that above.

Each of the formulas (6) and (7) expresses a relation among certain three parts of triangle ABC. Hence, given two of those parts, the third can be determined by using (6) or (7). Similarly, there is a formula relating each set of three parts (excluding the right angle), such as a, b, c, or λ, μ, a, and so forth.* Accordingly, there are eight other formulas besides (6) and (7). Some of them can be obtained by applying Theorem 4 to the triangles associated with triangle ABC, which were discussed in Section 2.

Consider, for example, the associated Triangle 2, whose parts are

$$m', \quad b, \quad l, \quad \gamma, \quad \alpha',$$

where m', b represent the arms, γ and α' the angles opposite them, respectively, and l the hypotenuse. Applying Theorem 4 to this triangle, we get

$$\sin \gamma = \frac{\sinh m'}{\sinh l}$$

and

$$\sin \alpha' = \frac{\sinh b}{\sinh l}.$$

These equations are easily reduced to formulas for triangle ABC by using the relations between distances and angles of parallelism (§3, Theo. 2). Thus, from the first of these equations we obtain

$$\operatorname{sech} c = \frac{\cot \mu'}{\cot \lambda} = \frac{\tan \mu}{\cot \lambda},$$

or

$$\cot \lambda \cot \mu = \cosh c, \tag{8}$$

and from the second we get (since $\sin \alpha' = \cos \alpha$)

$$\tanh a = \frac{\sinh b}{\cot \lambda},$$

or

$$\tan \lambda = \frac{\tanh a}{\sinh b}. \tag{9}$$

---

* This goes hand in hand with the fact that two right triangles are congruent if two parts (other than the right angle) of one triangle are equal, respectively, to the two analogous parts of the other.

Equation (8) shows how the acute angles and the hypotenuse of triangle ABC are related, while (9) expresses the relation between the arms and one acute angle. The relation between the arms and the other acute angle must be analogous to (9), that is,

$$\tan \mu = \frac{\tanh b}{\sinh a}. \tag{10}$$

We obtained formulas (8) through (10) by applying Theorem 4 to Triangle 2. No additional formulas for triangle ABC would result from applying this theorem to the other associated triangles. But if (9) and (10) are substituted in (8), a new formula for triangle ABC is obtained, as we shall presently see, and when this formula is applied to the associated triangles the remaining four formulas for triangle ABC will result. Making the specified substitution, we get

$$\frac{\sinh b}{\tanh a} \cdot \frac{\sinh a}{\tanh b} = \cosh c,$$

or

$$\cosh a \cosh b = \cosh c. \tag{11}$$

This formula gives the relation among the three sides of triangle ABC and is therefore analogous to the Pythagorean formula of Euclidean geometry. It will be useful to state (11) in words:

Theorem 5. *The hyperbolic cosine of the hypotenuse of a right triangle is equal to the product of the hyperbolic cosines of the other sides.*

Applying this to associated Triangle 2, whose parts are

$$m', \quad b, \quad l, \quad \gamma, \quad \alpha',$$

we get

$$\cosh m' \cosh b = \cosh l,$$

which, by Theorem 2, becomes

$$\csc \mu' \cosh b = \csc \lambda$$

and hence

$$\sec \mu \cosh b = \csc \lambda,$$

or

$$\cosh b = \frac{\cos \mu}{\sin \lambda}. \tag{12}$$

This formula relates the angles λ, μ and the side opposite μ in triangle ABC. The formula relating λ, μ, and the side opposite λ must then be

$$\cosh a = \frac{\cos \lambda}{\sin \mu}. \tag{13}$$

We leave it to the student to show that application of Theorem 5 to associated Triangle 4 leads to the formula

$$\cos \lambda = \frac{\tanh b}{\tanh c}. \tag{14}$$

The relations between complementary distances (§3, Theo. 3) will be found useful in doing this.

Formula (14) relates angle λ to the adjacent side and the hypotenuse, and hence corresponds to the Euclidean formula $\cos \lambda = b/c$. The relation between angle μ, the side adjacent to it, and the hypotenuse must be of the same nature as (14), and hence is

$$\cos \mu = \frac{\tanh a}{\tanh c}. \tag{15}$$

This completes the derivation of the right triangle formulas. Looking back, we see that formula (6) was the key to the others. We summarize the results:

Theorem 6. *If ABC is any right triangle, with sides and acute angles labeled a, b, c, λ, μ in the usual way, then the following formulas hold:*

Two Sides and an Angle

$$\sin \lambda = \frac{\sinh a}{\sinh c}, \qquad \cos \lambda = \frac{\tanh b}{\tanh c}, \qquad \tan \lambda = \frac{\tanh a}{\sinh b},$$

$$\sin \mu = \frac{\sinh b}{\sinh c}, \qquad \cos \mu = \frac{\tanh a}{\tanh c}, \qquad \tan \mu = \frac{\tanh b}{\sinh a}.$$

Two Angles and a Side

$$\cosh a = \frac{\cos \lambda}{\sin \mu}, \qquad \cosh b = \frac{\cos \mu}{\sin \lambda}, \qquad \cosh c = \cot \lambda \cot \mu.$$

Three Sides

$$\cosh a \cosh b = \cosh c.$$

The exercises will give the student an opportunity to work with these formulas and perhaps to understand them better. In the next section we discuss their relation to the corresponding Euclidean formulas.

EXERCISES

1. Find $\sin \lambda$ and λ (nearest degree) for the following right triangles:

(a) $a = 1, c = 2$; (b) $a = 2, c = 4$; (c) $a = 0.5, c = 1$.

2. Find $\cos \lambda$ and λ (nearest degree) for the following right triangles:

(a) $b = 1, c = 2$; (b) $b = 2, c = 4$; (c) $b = 0.5, c = 1$.

3. Find $\tan \lambda$ and λ (nearest degree) for the following right triangles:

(a) $a = b = 1$; (b) $a = b = 2$; (c) $a = b = 0.5$.

4. Find the hypotenuse in each of the following right triangles:

(a) $a = 1, b = 1$; (b) $a = 1, b = 2$; (c) $a = 3, b = 4$.

5. Solve the following right triangles:

(a) $\lambda = \mu = 30°$; (b) $\lambda = \mu = 40°$; (c) $a = 3, c = 5$.

6. Formula (6) was proved only for the case $l > c$. Prove it when (a) $l = c$, (b) $l < c$.

7. Prove that $\sinh(A - B) = \sinh A \cosh B - \cosh A \sinh B$, where A, B are any positive numbers and $A > B$.*

8. Find the two formulas for triangle ABC which result from applying Theorem 4 to associated Triangle 3.

9. Show that formula (14) can be obtained by applying Theorem 5 to associated Triangle 4.

10. Find a formula for triangle ABC by applying Theorem 5 to associated Triangle 3.

11. If a, b, c are the lengths of the arms, base, and summit of a Saccheri quadrilateral, prove that the length d of the segment joining the midpoints of the base and summit is given by the formula

$$\cosh d = \frac{\cosh a \cosh(b/2)}{\cosh(c/2)}.$$

* Actually, the formula holds for all values of A and B.

6. COMPARISON WITH EUCLIDEAN FORMULAS

Each of the right triangle formulas derived in the preceding section *looks* different from the corresponding Euclidean formula, and therefore presumably *is* different, that is, generally gives different results. The preceding exercises support this view. They permit us to conclude, for example, in the case of the formula sin λ = sinh a/sinh c, that sinh a/sinh c cannot be simplified to a/c. A question that still remains, though, is whether these two ratios might be equal for *some* values of a and c. If they could, then the Euclidean and hyperbolic formulas for sin λ would be equivalent for those values. We shall now show that the two ratios are never equal.

It is an established fact that, for all values of x, the hyperbolic function sinh x can be expressed as an infinite series in the following way:

$$\sinh x = x + \frac{x^3}{3!} + \frac{x^5}{5!} + \frac{x^7}{7!} + \cdots .$$

Hence, when $x = a$ and $x = c$, we have

$$\sinh a = a + \frac{a^3}{3!} + \frac{a^5}{5!} + \frac{a^7}{7!} + \cdots ,$$

$$\sinh c = c + \frac{c^3}{3!} + \frac{c^5}{5!} + \frac{c^7}{7!} + \cdots .$$

Both sides of these equations may be divided by the same number (not zero) just as if the infinite expressions on the right were polynomials. Dividing the first by a, and the second by c, we get

$$\frac{\sinh a}{a} = 1 + \frac{a^2}{3!} + \frac{a^4}{5!} + \frac{a^6}{7!} + \cdots , \tag{1}$$

$$\frac{\sinh c}{c} = 1 + \frac{c^2}{3!} + \frac{c^4}{5!} + \frac{c^6}{7!} + \cdots . \tag{2}$$

Since $c > a$, each term on the right side of (2) other than the first exceeds the corresponding term of (1). Hence the entire right side of (2) exceeds the entire right side of (1). The same being true of the left sides, we have

$$\frac{\sinh a}{a} < \frac{\sinh c}{c},$$

or

$$\frac{\sinh a}{\sinh c} < \frac{a}{c}.$$

Thus the ratios sinh a/sinh c and a/c are never equal.

Since $\sin \lambda = \sinh a/\sinh c$, we see that in hyperbolic geometry the sine of an acute angle in a right triangle is never equal to the opposite side divided by the hypotenuse. Corresponding statements can be made for the other trigonometric functions. For example, the cosine of an acute angle in a right triangle never equals the adjacent side divided by the hypotenuse. We shall prove this later by showing that $\tanh b/\tanh c$, which equals $\cos \lambda$, always exceeds b/c. Using this fact now, however, we can easily see that the tangent never equals the opposite side divided by the adjacent side. For, since $\sin \lambda < a/c$ and $\cos \lambda > b/c$, we can write

$$\tan \lambda = \frac{\sin \lambda}{\cos \lambda} < \frac{a/c}{b/c} = \frac{a}{b},$$

or

$$\tan \lambda < \frac{a}{b}.$$

In summary we have

Theorem 7. *If λ is an acute angle in a right triangle labeled in the usual way, then*

$$\sin \lambda < \frac{a}{c}, \qquad \cos \lambda > \frac{b}{c}, \qquad \tan \lambda < \frac{a}{b},$$

and corresponding inequalities hold for $\csc \lambda$, $\sec \lambda$, $\cot \lambda$. *The values of these six trigonometric functions of λ are therefore never equal to the ratios of the sides of a right triangle, as they are in Euclidean geometry.*

In all Euclidean right triangles having an acute angle of given size λ we know that each specified ratio of two sides, such as a/c, b/c, a/b, and so on, has a constant value. Theorem 7 does not say that this cannot also occur in hyperbolic geometry. If it does occur, however, the theorem shows that those constant values are not $\sin \lambda$, $\cos \lambda$, $\tan \lambda$, and so forth. Actually, it does not occur. We shall now prove this by showing that each ratio cannot have the same value even twice, that is, the same value in two noncongruent triangles.

Theorem 8. *Let λ be an acute angle of any right triangle labeled in the usual way. If λ remains fixed and c increases, then each ratio of two sides either always increases or always decreases, and hence does not have the same value twice. In particular, the ratios a/c and a/b always increase, whereas the ratio b/c always decreases.*

Proof. We shall give the proof only for the ratios a/c, b/c, a/b. Consider the right triangle formula $\sin \lambda = \sinh a/\sinh c$, or

$$\sinh a = \sin \lambda \sinh c. \tag{1}$$

Let c increase. Then $\sinh c$ increases (Ex. 13). Since λ remains fixed, we see from (1) that $\sinh a$ increases. Hence a increases (Ex. 13). To show that the ratio a/c also increases we shall consider the more general equation

$$\sinh y = k \sinh x, \tag{2}$$

where k is any positive constant less than 1 and x, y are nonnegative variables, and show that y/x always increases when x does.

As a table of values would suggest (Ex. 5), and as can be proved by elementary calculus, the ordinary (Euclidean) graph of (2) goes through the origin O and always rises at an increasing rate when x increases. It is therefore concave upward everywhere and looks as shown in Fig. VI, 6. If $P_1(x_1, y_1)$,

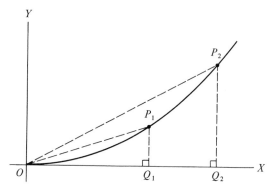

Fig. VI, 6

$P_2(x_2, y_2)$ are any two points on the graph such that $x_2 > x_1 > 0$, and Q_1, Q_2 are their projections on the X-axis, then

$$\angle P_2OQ_2 > \angle P_1OQ_1$$

$$\tan \angle POQ_2 > \tan \angle P_1OQ_1$$

$$\frac{Q_2P_2}{OQ_2} > \frac{Q_1P_1}{OQ_1}$$

$$\frac{y_2}{x_2} > \frac{y_1}{x_1}.$$

Thus, y/x always increases when x increases. This proves the part of Theorem 8 dealing with the ratio a/c.

The proof of the part for the ratio b/c is similar, but we now use the right triangle formula $\cos \lambda = \tanh b/\tanh c$, or

$$\tanh b = \cos \lambda \tanh c. \tag{3}$$

Let c increase. Then $\tanh c$ increases (this can be checked by a table of values or proved by elementary calculus). Since λ remains fixed, (3) shows that $\tanh b$ increases. Hence b increases. To see that the ratio b/c decreases we consider the equation

$$\tanh y = k \tanh x, \tag{4}$$

where k, x, y have the same meanings as in (2), and show that y/x always decreases when x increases. The graph of (4) goes through the origin and always rises at a diminishing rate as x increases. It is therefore concave downward everywhere (Fig. VI, 7). The remaining details, being much like those used earlier for the ratio a/c, are left as an exercise.

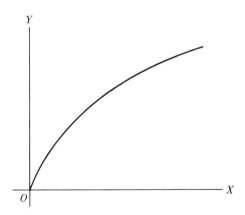

Fig. VI, 7

The ratio a/b is now easily shown to increase by writing

$$\frac{a}{b} = \frac{a}{c} \div \frac{b}{c}$$

and using what we know about a/c and b/c.

Our discussion thus far has drawn attention only to the ways in which the right triangle formulas of hyperbolic geometry differ from those of Euclidean geometry. However, there is also a close relation between the two sets of formulas. When the sides of a right triangle are very small, each hyperbolic formula is approximately equivalent to the corresponding Euclidean formula. In the case of the formulas involving two sides and an angle we have the following more precise statement.

Theorem 9. *Let λ be an acute angle in any right triangle labeled in the usual way. If λ remains fixed and c approaches zero, the ratios a/c, b/c, a/b approach the constant values $\sin \lambda$, $\cos \lambda$, $\tan \lambda$. Hence, when c is very small, the ratios are approximately equal to these values, or, in symbols,*

$$\sin \lambda \approx \frac{a}{c}, \qquad \cos \lambda \approx \frac{b}{c}, \qquad \tan \lambda \approx \frac{a}{b},$$

the approximations generally improving as c gets smaller.

For example, in a right triangle with the comparatively small measures $a = 0.1$, $c = 0.2$, we have

$$\sin \lambda = \frac{\sinh a}{\sinh c} = \frac{\sinh 0.1}{\sinh 0.2} = \frac{0.10017}{0.20134} \approx \frac{0.1}{0.2}.$$

Thus, $\sin \lambda$ is very close to $\frac{1}{2}$, the value it would have in a Euclidean right triangle with $a = 0.1$, $c = 0.2$.

Proof of Theorem 9. We shall make use of the fact that $\cosh x$ and $(\sinh x)/x$ both approach 1 when x approaches 0. That this is reasonable can be seen from inspection of the series

$$\cosh x = 1 + \frac{x^2}{2!} + \frac{x^4}{4!} + \frac{x^6}{6!} + \cdots ,$$

$$\frac{\sinh x}{x} = 1 + \frac{x^2}{3!} + \frac{x^4}{5!} + \frac{x^6}{7!} + \cdots ,$$

where x is any positive number. In the case of $\cosh x$ it can also be seen by consideration of the definition $\cosh x = (e^x + e^{-x})/2$.

We first prove that a/c approaches $\sin \lambda$. Since

$$\sin \lambda = \frac{\sinh a}{\sinh c},$$

we can write

$$\sin \lambda = \frac{a}{c} \cdot \frac{\dfrac{\sinh a}{a}}{\dfrac{\sinh c}{c}},$$

or

$$\frac{a}{c} = \sin \lambda \, \frac{\dfrac{\sinh c}{c}}{\dfrac{\sinh a}{a}}. \tag{1}$$

Now, keeping λ fixed, let c approach 0. Then a, being less than c, also approaches 0. Since $(\sinh a)/a$ and $(\sinh c)/c$ will then both approach 1, while $\sin \lambda$ remains fixed, the right side of (1) approaches the value $\sin \lambda$. The left side must therefore do likewise.

To prove that b/c approaches $\cos \lambda$, we write

$$\cos \lambda = \frac{\tanh b}{\tanh c} = \frac{\dfrac{\sinh b}{\cosh b}}{\dfrac{\sinh c}{\cosh c}} = \frac{\sinh b}{\sinh c} \cdot \frac{\cosh c}{\cosh b}$$

$$= \frac{b}{c} \cdot \frac{\dfrac{\sinh b}{b}}{\dfrac{\sinh c}{c}} \cdot \frac{\cosh c}{\cosh b},$$

so that

$$\frac{b}{c} = \cos \lambda \, \frac{\dfrac{\sinh c}{c}}{\dfrac{\sinh b}{b}} \cdot \frac{\cosh b}{\cosh c}. \tag{2}$$

When c approaches 0, so does b since it is less than c. Hence $\cosh b$, $\cosh c$ approach 1 and the right side of (2) approaches the fixed value $\cos \lambda$. The left side must therefore do likewise.

The proof that a/b approaches $\tan \lambda$ is left as an exercise.

We can now return to the proof of Theorem 7 and show that $\cos \lambda > b/c$. (This part of the proof had been postponed.) By Theorem 8, b/c decreases when c increases, λ remaining fixed. It therefore increases when c decreases. In particular, if c decreases so as to approach 0, then b/c will increase and, by Theorem 9, approach $\cos \lambda$. Therefore b/c must be less than $\cos \lambda$, or $\cos \lambda > b/c$.

Theorem 9 applies only to the right triangle formulas involving two sides and an acute angle. The facts concerning the formulas involving only one side are brought out in exercises. The remaining formula

$$\cosh c = \cosh a \cosh b \tag{3}$$

will be considered now and shown to be approximately equivalent to the Euclidean formula $c^2 = a^2 + b^2$ when c is very small.

Using the infinite series for $\cosh x$ given earlier, we can write (3) as

$$1 + \frac{c^2}{2!} + \frac{c^4}{4!} + \frac{c^6}{6!} + \cdots$$

$$= \left(1 + \frac{a^2}{2!} + \frac{a^4}{4!} + \frac{a^6}{6!} + \cdots\right)\left(1 + \frac{b^2}{2!} + \frac{b^4}{4!} + \frac{b^6}{6!} + \cdots\right). \tag{4}$$

It is permissible to multiply the two series on the right as we would multiply two polynomials and thus obtain the series representing their product. Equation (4) then becomes

$$1 + \frac{c^2}{2!} + \frac{c^4}{4!} + \frac{c^6}{6!} + \cdots = 1 + \frac{a^2}{2!} + \frac{b^2}{2!} + \frac{a^4}{4!} + \frac{a^2 b^2}{2!2!} + \frac{b^4}{4!} + \frac{a^6}{6!}$$

$$+ \frac{a^4 b^2}{4!2!} + \frac{a^2 b^4}{2!4!} + \frac{b^6}{6!} + \cdots. \tag{5}$$

When c is very small (say 0.01 or less), so are a and b. Then, on each side of (5) the terms of fourth, sixth, and higher degrees have values very much less than those of the other terms, and it can be shown that their sum is likewise very much less than the sum of the other terms. The left side is therefore approximately equal to the sum of its first two terms and the right side to the sum of its first three terms. Thus, (5) is approximately equivalent to the equation

$$1 + \frac{c^2}{2!} = 1 + \frac{a^2}{2!} + \frac{b^2}{2!},$$

or

$$c^2 = a^2 + b^2,$$

and, in general, the smaller c is, the better the approximation.

EXERCISES

1. Give the explicit statements for $\csc \lambda$, $\sec \lambda$, $\cot \lambda$ in Theorem 7.

2. Theorem 8 does not specifically mention the ratios c/a, c/b, b/a. How do they vary?

3. Show that the formula $\cosh a = \cos \lambda / \sin \mu$ is never equivalent to the corresponding Euclidean formula $\sin \mu = \cos \lambda$, but approximates it more and more closely as c approaches 0.

4. Which Euclidean formula can be regarded as corresponding to the formula $\cosh c = \cot \lambda \cot \mu$? (See Ex. 3.) Show that the two formulas are never equivalent, but become more nearly so as c approaches 0.

5. (a) Using a table of values, draw the graph of $\sinh y = \frac{1}{2} \sinh x$ from $x = 0$ to $x = 5$. (b) Calculate y/x from your table whenever possible and verify that it increases when x increases.

6. (a) Using a table of values, draw the graph of $\tanh y = \frac{1}{2} \tanh x$ from $x = 0$ to $x = 5$. (b) Calculate y/x from your table whenever possible and verify that it decreases when x increases.

7. Complete the proof that b/c decreases when c increases, λ remaining fixed.

8. Using the method suggested in the discussion, prove that a/b increases when c increases, λ remaining fixed.

9. Prove that a/b approaches $\tan \lambda$ when c approaches 0, λ remaining fixed.

10. Find c in each case and compare it with the Euclidean value:

(a) $a = 0.3$, $b = 0.4$;   (b) $a = 0.03$, $b = 0.04$.

11. In each case find λ to the nearest minute and compare it with the Euclidean value:

(a) $b = 0.1$, $c = 0.2$;   (b) $b = 0.05$, $c = 0.1$;
(c) $a = b = 0.1$;   (d) $a = b = 0.05$.

12. Find μ to the nearest minute in each case and compare it with the Euclidean value:

(a) $\lambda = 45°$, $c = 0.1$;   (b) $\lambda = 45°$, $c = 0.05$.

13. If x is positive and increases, prove that $\sinh x$ also increases. Using this fact, prove, conversely, that if $\sinh x$ increases, then so does x.

7. FORMULAS FOR THE GENERAL TRIANGLE

Let us now consider an oblique triangle, that is, a triangle containing no right angle, and determine the relations among its sides and angles. We shall limit ourselves to those relations which are analogous to the Euclidean Law of Sines and Law of Cosines. As in the Euclidean situation, the relations are found by applying right triangle formulas to certain right triangles that can be formed from the oblique triangle.

Let *ABC* (Fig. VI, 8) be any oblique triangle, with angles at *A*, *B*, *C* of size λ, μ, v and sides opposite these angles of length *a*, *b*, *c*. We shall consider only the case in which each angle is acute. If *D* is the projection of *C* on $\overline{AB}$, then *D* is between *A* and *B*, so that right triangles *ACD*, *BCD* are formed

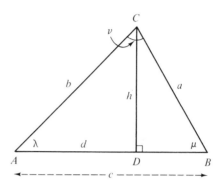

Fig. VI, 8

containing ⊀*A*, ⊀*B* of the original triangle. Applying right triangle formulas (§5, Theo. 6) to these triangles, we obtain

$$\sin \lambda = \frac{\sinh h}{\sinh b}, \qquad \sin \mu = \frac{\sinh h}{\sinh a}.$$

Hence

$$\frac{\sin \lambda}{\sin \mu} = \frac{\sinh a}{\sinh b},$$

or

$$\frac{\sinh a}{\sin \lambda} = \frac{\sinh b}{\sin \mu}. \tag{1}$$

Similarly, using the altitude from *A* to $\overline{BC}$ we obtain

$$\frac{\sinh b}{\sin \mu} = \frac{\sinh c}{\sin v}. \tag{2}$$

Combining (1) and (2) gives

$$\frac{\sinh a}{\sin \lambda} = \frac{\sinh b}{\sin \mu} = \frac{\sinh c}{\sin v}. \tag{3}$$

This formula expresses a relation among all the sides and angles of triangle ABC. Being analogous to the Euclidean Law of Sines

$$\frac{a}{\sin \lambda} = \frac{b}{\sin \mu} = \frac{c}{\sin \nu},$$

it may be called the *hyperbolic Law of Sines*.

Formula (3) also holds when an angle of triangle ABC is obtuse. The proof of this is left as an exercise. Moreover, the formula holds when ABC is a right triangle. For example, if $\nu = 90°$, (3) reduces to

$$\sin \lambda = \frac{\sinh a}{\sinh c} \quad \text{and} \quad \sin \mu = \frac{\sinh b}{\sinh c},$$

which are correct right triangle relations. Thus formula (3) is true not only for oblique triangles but for right triangles as well. It is therefore a formula for the *general triangle*.

Returning now to triangle ABC (Fig. VI, 8), let us derive another relation, again assuming that λ, μ, ν are acute. By Theorem 6, Section 5, we have the following equations for the sides of right triangles ACD and BCD:

$$\cosh a = \cosh h \cosh(c - d)$$

$$\cosh b = \cosh h \cosh d.$$

Hence

$$\frac{\cosh a}{\cosh b} = \frac{\cosh(c - d)}{\cosh d}$$

$$\cosh a = \frac{\cosh b \cosh(c - d)}{\cosh d}$$

$$= \frac{\cosh b(\cosh c \cosh d - \sinh c \sinh d)^*}{\cosh d}$$

$$= \cosh b \cosh c - \cosh b \sinh c \tanh d. \tag{4}$$

Since $\cos \lambda = \tanh d/\tanh b$, we have $\tanh d = \tanh b \cos \lambda$. Substituting this in (4), we get

$$\cosh a = \cosh b \cosh c - \sinh b \sinh c \cos \lambda. \tag{5}$$

This formula relates the three sides and one angle of triangle ABC. It therefore corresponds to the Euclidean Law of Cosines

$$a^2 = b^2 + c^2 - 2bc \cos \lambda \tag{6}$$

* The proof of the formula for $\cosh(c - d)$ used here is considered in an exercise.

and may be called the *hyperbolic Law of Cosines*. Analogous to (5) there is a formula involving μ, and another involving v. It is left as an exercise to derive (5) when one angle of triangle ABC is obtuse. Also, the student can show that (5) is true when one angle is a right triangle. Thus (5) is a formula for the general triangle.

Our results can now be summarized.

Theorem 10.   *If a, b, c are the lengths of the sides of any triangle, and λ, μ, v are the measures of the opposite angles, respectively, then*

$$\frac{\sinh a}{\sin \lambda} = \frac{\sinh b}{\sin \mu} = \frac{\sinh c}{\sin v},$$

$$\cosh a = \cosh b \cosh c - \sinh b \sinh c \cos \lambda,$$

and formulas analogous to the latter hold for μ and v.

As can be expected, computations based on the hyperbolic Law of Sines and the hyperbolic Law of Cosines give different results from those based on their Euclidean counterparts. This is checked in the exercises. When the sides of the triangle are very small, however, it will be found that the two sets of results are in close agreement. This, too, is not surprising inasmuch as the two hyperbolic laws for the general triangle were deduced from right triangle formulas. Actually, it can be proved that the two laws become more nearly equivalent to their Euclidean counterparts when the sides of the triangle are taken shorter and shorter. The proof in the case of the hyperbolic Law of Sines should present no difficulty to the student in view of what was done in the preceding section and hence is left as an exercise. The proof in the case of the hyperbolic Law of Cosines is as follows.

Let us suppose that $a > b$, $a > c$ and that a is very small. Then b, c are also very small. Using infinite series for $\cosh a$, $\cosh b$, $\cosh c$ and reasoning as in the preceding section, we know that

$$\cosh a \approx 1 + \frac{a^2}{2!} \quad \text{and} \quad \cosh b \cosh c \approx 1 + \frac{b^2}{2!} + \frac{c^2}{2!}. \tag{7}$$

Then, from the fact that

$$\sinh b \sinh c \cos \lambda = bc \, \frac{\sinh b}{b} \cdot \frac{\sinh c}{c} \cos \lambda$$

and

$$\frac{\sinh b}{b} \approx 1, \qquad \frac{\sinh c}{c} \approx 1,$$

we also know that

$$\sinh b \sinh c \cos \lambda \approx bc \cos \lambda. \tag{8}$$

Substituting (7) and (8) in (5) gives

$$1 + \frac{a^2}{2!} \approx 1 + \frac{b^2}{2!} + \frac{c^2}{2!} - bc \cos \lambda,$$

or

$$a^2 \approx b^2 + c^2 - 2bc \cos \lambda.$$

In other words, if a, b, c are very small and satisfy the hyperbolic Law of Cosines (5), then they also approximately satisfy the Euclidean Law of Cosines (6). This approximation is generally better, the smaller a, b, c are, since the same is true of the approximations (7) and (8).

EXERCISES

1. Find the angles (to the nearest minute) in the equilateral triangles whose sides have the following lengths:

(a) 1; (b) 0.5; (c) 0.2; (d) 0.1.

2. In each of the following oblique triangles $\lambda = 30°$ and $\mu = 60°$. Find b for the given value of a and compare it with the corresponding Euclidean value:

(a) $a = 2$; (b) $a = 1$; (c) $a = 0.3$; (d) $a = 0.1$.

3. Derive the hyperbolic Law of Sines when $\lambda > 90°$.

4. Derive the hyperbolic Law of Cosines (5) when $\lambda > 90°$.

5. Show that the hyperbolic Law of Cosines (5) holds when $\lambda = 90°$.

6. Prove that the hyperbolic Law of Sines becomes more nearly equivalent to the Euclidean Law of Sines as a, b, c approach 0.

7. If a Saccheri quadrilateral has arms of length a and a base of length b, show that the length c of its summit is given by the formula

$$\cosh c = \cosh^2 a \cosh b - \sinh^2 a.$$

8. If a, b, c are the lengths of the arms, base, and summit of a Saccheri quadrilateral, prove that each summit angle θ is given by the equation

$$\cos \theta = \frac{\sinh a \cosh a(\cosh b - 1)}{\sinh c}.$$

The formula of Exercise 7 may be used.

9. Prove the formula

$$\cosh(c - d) = \cosh c \cosh d - \sinh c \sinh d.$$

8. HYPERBOLIC GEOMETRY IN SMALL REGIONS

In an earlier chapter attention was called to the fact that locally, or in the small, the hyperbolic plane is approximately Euclidean. By this we meant that the measures of very small figures in hyperbolic geometry fit the formulas of Euclidean geometry very closely (though not exactly) and that any desired precision of fit can be obtained by taking the figures sufficiently small. Two illustrations of this fact were given. First, we saw that the angle-sum of a triangle approaches 180° as the area approaches 0, and vice versa. Then we noted that the summit angles of a Saccheri quadrilateral approach the Euclidean value of 90° when the area of the quadrilateral approaches 0, and vice versa. Our work in the present chapter has provided us with a third illustration, this time involving length rather than area. We saw that the formulas relating the sides and angles of any triangle are approximately equivalent to the corresponding Euclidean formulas when the sides are very small, the approximation generally improving as the sides get smaller. After obtaining some immediate consequences of the preceding results we shall further illustrate the fact that, locally, the hyperbolic plane is approximately Euclidean.

Since the hyperbolic triangle formulas become more nearly Euclidean as the sides approach 0 and the angle-sum of a Euclidean triangle is 180°, it follows that the angle-sum of a hyperbolic triangle must approach 180° as the sides approach 0. But we know that the area of the triangle approaches 0 as the angle-sum approaches 180°. Consequently, triangles whose sides are very small also have areas that are very small. This was not known from the discussion of area in Chapter III since we had no formula relating the area of a triangle to the lengths of the sides.

Having already seen that the summit angles of a Saccheri quadrilateral approach the Euclidean value of 90° when the area of the quadrilateral approaches 0, we shall now show that they do likewise when the sides approach 0. Let $ABCD$ (Fig. VI, 9) be any Saccheri quadrilateral, with right angles at A, B, and let $\overline{EF}$ be perpendicular to the base and summit. When the sides of the quadrilateral are very small, so is $\overline{AF}$, and also $\overline{EF}$ since $EF < AD$. Then $\overline{AE}$, the hypotenuse of triangle AEF, is very small. For, from the equation $\cosh(AE) = \cosh(AF)\cosh(EF)$ we see that $\cosh(AE) \approx 1$, and

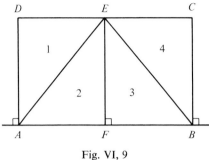

Fig. VI, 9

hence that $AE \approx 0$. Similarly, $\overline{BE}$ is very small. Thus all the sides of triangles 1 through 4 are very small. Hence the angle-sums of these triangles are close to 180°, the total of these sums is close to 720°, the angle-sum of the quadrilateral is close to 360°, and hence each summit angle is close to 90°. Clearly, each summit angle can be brought arbitrarily close to 90° by making the sides of the quadrilateral sufficiently small.

For our next illustration of the fact that hyperbolic geometry is, in the small, approximately Euclidean, we shall consider the relation between a distance x and the corresponding angle of parallelism y:

$$\cos y = \tanh x.$$

Let x approach 0. Then $\tanh x$ approaches 0, as is easily seen from the definition $\tanh x = \sinh x/\cosh x$. Hence, from the equation above, $\cos y$ approaches 0, so that y approaches 90°. Although we have had no occasion to define an angle of parallelism in Euclidean geometry, its value, had we done so, would clearly always be 90°. To interpret geometrically the approach of y to 90°, let m, n (Fig. VI, 10) be the asymptotic parallels to g through P, let x be the distance PQ from P to g, and let h be the horizontal line through P, that is, the line through P perpendicular to $\overline{PQ}$. Keeping g and Q fixed, let P approach

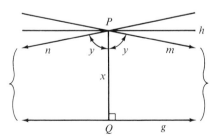

Fig. VI, 10

Q along $\overline{PQ}$. As x thus approaches 0, the angle of parallelism y approaches 90°, with the result that m, n, and each nonasymptotic parallel between them become more nearly horizontal. In other words, all the parallels to g through P tend to coalesce into a single horizontal parallel, which is the Euclidean situation.

Finally, let us consider circles. It can be shown that the formula relating the circumference and radius of a circle is

$$C = 2\pi \sinh r \tag{1}$$

when the standard unit of length is used. From the series

$$\sinh r = r + \frac{r^3}{3!} + \frac{r^5}{5!} + \cdots$$

we see that $\sinh r > r$. Therefore, from (1) it follows that $C > 2\pi r$. In other words, when the standard unit of length is used, the circumference of a circle of given radius is always greater than in Euclidean geometry. (The Euclidean formula, of course, is the same regardless of the unit of length.) However, the hyperbolic formula becomes more nearly equivalent to the Euclidean formula as r approaches 0. This is shown by writing (1) in the form

$$C = 2\pi r \frac{\sinh r}{r}$$

and recalling from Section 6 that $(\sinh x)/x$ approaches 1 when x approaches 0.

It can be proved that the area of a circle is related to its radius by the formula

$$A = 4\pi \sinh^2 \frac{r}{2} \tag{2}$$

when the standard unit of length is used. The region serving as unit of area in this formula is shown in Fig. VI, 11, where $A\delta$, $B\delta$ are boundary rays and

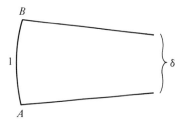

Fig. VI, 11

$\overset{\frown}{AB}$ is an arc of unit length of a horocycle with direction δ. Writing (2) in the form

$$A = \pi r^2 \left(\frac{\sinh(r/2)}{r/2}\right)^2 \tag{3}$$

and recalling from Section 6 that

$$\frac{\sinh x}{x} = 1 + \frac{x^3}{3!} + \frac{x^5}{5!} + \cdots$$

for all positive values of x, we see that

$$\frac{\sinh(r/2)}{r/2} > 1$$

and hence that $A > \pi r^2$. Thus, when the standard unit of length is used, the area of a circle of given radius is always greater than in Euclidean geometry. (The Euclidean area formula is the same regardless of the unit of length.) However, it is easily seen that the hyperbolic formula becomes more nearly equivalent to the Euclidean formula when the radius becomes smaller inasmuch as the expression within the large parentheses in (3) approaches 1 when r approaches 0.

We have used the statement *hyperbolic geometry, in the small, or locally, is approximately Euclidean* to mean that the measures of very small figures in hyperbolic geometry fit Euclidean formulas very closely (though not exactly), any desired precision of fit being obtainable by taking the figures sufficiently small. In our discussion "small figure" has sometimes meant a figure of small area, sometimes a figure of small linear dimensions, that is, with small sides, radius, circumference, and so on. While it is true, for example, that a triangle with very small sides will also have a very small area, a triangle of very small area clearly need not have sides which are all very small (consider, for example, a long, very slender triangle). To avoid this ambiguity we can introduce the term *small region* to mean a portion of the plane in which all distances are small (that is, less than some small number) and agree that a *small figure* is any figure in such a region. The italicized statement at the beginning of this paragraph can then be interpreted to mean that the measures of figures in very small regions of the hyperbolic plane closely obey Euclidean formulas, any desired degree of closeness being obtainable by taking the region sufficiently small (that is, so that all distances in it are less than some sufficiently small number).

EXERCISES

1. Give an argument showing that Saccheri quadrilaterals with very small sides also have very small areas.

2. Show that the fourth angle of a Lambert quadrilateral approaches 90° when the sides approach 0.

3. Prove that the ratio of the circumference of a circle to the diameter (a) is not constant, (b) exceeds π, (c) approaches π when the diameter approaches 0.

4. Calculate the circumference of a circle for the following values of the radius and compare the results with the corresponding Euclidean values:

(a) 1.5; (b) 1; (c) 0.5; (d) 0.25.

5. Calculate the area of a circle for the following values of the radius and compare the results with the corresponding Euclidean values:

(a) 2; (b) 1; (c) 0.5; (d) 0.2.

6. Show that a region (boundary included) in which all distances do not exceed a specified positive value d can be obtained by taking the boundary to be a circle of radius $d/2$.

7. Show that a quadrilateral with equal sides and equal angles can be obtained by "putting together" four congruent isosceles right triangles and that it becomes more nearly a square as its sides approach 0.

8. Prove algebraically that when the arms of a Saccheri quadrilateral approach 0, the base remaining constant, (a) the length of the summit approaches the length of the base (use §7, Ex. 7), and (b) each summit angle approaches 90° [use part (a) and §7, Ex. 8].

9. Saccheri quadrilaterals whose sides are not small may differ considerably from their Euclidean counterparts. Verify this by showing that when the arms become arbitrarily great, the base remaining constant, (a) the length of the summit increases beyond all bounds (use §7, Ex. 7), and (b) the summit angles approach 0° (use §7, Ex. 8).

9. HYPERBOLIC GEOMETRY AND THE PHYSICAL WORLD

Our work in the preceding sections suggests that the formulas of hyperbolic geometry, being approximately equivalent to the corresponding Euclidean formulas when the measures involved are small, ought to give results in close agreement with physical experience if they are applied to sufficiently small physical regions. The question then arises: When is a physical region sufficiently small in order that such agreement will occur? Conceivably, the region would have to be microscopic, in which case hyperbolic geometry might be of quite limited applicability. On the other hand, since "small" is a relative term, "sufficiently small" might turn out to mean merely small in relation to the earth's radius, or to the distance from the earth to the sun, or to some other very considerable distance occurring in astronomical work. In that case hyperbolic geometry would be widely applicable in science and technology, just as Euclidean geometry is.

It is common knowledge that Euclidean geometry applies very well to the physical world of experience and is therefore extensively used in science, engineering, and many other fields dealing with geometrical concepts. Such applicability is easy to understand when we realize that the statements on which all of Euclidean geometry rests, the axioms, themselves appear to be in close agreement with that world. This being so, hyperbolic geometry should be even more applicable than Euclidean geometry, for the axioms of the two systems are the same except for their parallel axioms and, as we shall now see, the hyperbolic parallel axiom appears to agree even better with experience than does the Euclidean parallel axiom.

In verifying this let us take as the hyperbolic parallel axiom, not the statement of it that we have been using, but the equivalent statement: *Given any line and any point not on it, there exists more than one line through the point which does not meet the given line.* Keeping in mind that the infinite extent of the physical universe is only an assumption and that our experience is always limited to a finite region, consider any plane region of experience *R* (Fig. VI, 12), any physical straight line *g* in *R*, and any physical point *P* in *R*

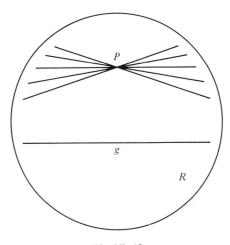

Fig. VI, 12

but not on g. It is a fact of experience that through P there pass not just one but many physical straight lines which lie in R and do not meet g however far they and g are extended in R, and that this is true regardless of the size of R. If R is a sheet of paper, for example, we all know that many pencil lines not meeting g can be ruled through P, and that this is true whether the sheet is small or large. Thus the hyperbolic parallel axiom appears to fit experience better than does the Euclidean parallel axiom.

Ordinarily we regard flat surfaces as being physical models of parts of Euclidean planes, the straight paths on those surfaces as being physical models of parts of Euclidean straight lines, and the various configurations on those surfaces as being physical models of Euclidean figures. Since hyperbolic geometry appears to fit experience at least as well as Euclidean geometry, these very same flat surfaces, straight paths, and configurations can, with equal correctness, be regarded as physical models of hyperbolic planes, hyperbolic straight lines, and hyperbolic figures. The angle-sum of a physical triangle can therefore be thought of as less than 180°, the summit angles of a physical Saccheri quadrilateral as acute, the ratio of the circumference of a physical circle to its diameter as exceeding π, and so forth, provided that one is careful not to say how much less than 180°, how acute, how much in excess of π, and so on. This attitude is legitimate regardless of the size of the physical figure. Since, as we have seen, the axioms of hyperbolic geometry appear to fit experience better than do those of Euclidean geometry, it would seem that very careful measurements of physical triangles, quadrilaterals, circles, and so forth, might give results more in conformity with hyperbolic formulas than with Euclidean ones. However, such results have not been obtained. This means that the values given by hyperbolic formulas when applied to physical figures are so close to those given by Euclidean formulas that the difference cannot be detected by physical measurements.

In the light of what was shown in the preceding section, then, we can conclude that the geometric figures dealt with in empirical work, even those whose size is ordinarily regarded as being very great, are all very small when expressed in terms of the standard unit of length. This unit must therefore represent a very great number of miles when interpreted physically. To improve our appreciation of this fact, which was already noted in Chapter IV, Section 6, let us recall that the standard unit is the segment corresponding to an angle of parallelism of about 40°24′ and then consider the physical triangle ABC shown in Fig. VI, 13. Here A represents the earth, C the sun, and B the comparatively close star Alpha Centauri. Astronomers have measured $\measuredangle A$ and found it to be 89°59′59″, to the nearest second. (Of course, no attempt has been made to draw Fig. VI, 13 with anything like this accuracy.) Since any physical triangle may be viewed from the standpoint of hyperbolic theory, we may say that the angle of parallelism corresponding to $\overline{AC}$ exceeds

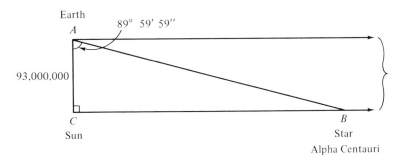

Fig. VI, 13

89°59'59". Hence the segment corresponding to an angle of parallelism less than 89°59'59" will exceed $\overline{AC}$, whose length is about 93 million miles. The angle 40°24' being very much less than 89°59'59", the segment corresponding to it, that is, the standard unit segment, will greatly exceed $\overline{AC}$. It can be shown, still using this same star, that the unit segment is actually more than 100,000 times $\overline{AC}$, and hence has a length exceeding 9 trillion miles. When more remote stars are used, the unit segment turns out to be even greater than this by far, but its precise extent has never been determined.

We can now see that the dimensions of any geometric figure likely to be considered in scientific work would be extremely small when expressed in terms of the standard unit of length and that any results for this figure obtained by hyperbolic computations would therefore be indistinguishable from those given by Euclidean computations. The following table will help to make this more concrete. It shows the result of calculating the hypotenuse c of a right triangle by the hyperbolic formula $\cosh c = \cosh a \cosh b$ and also by the Euclidean formula $c^2 = a^2 + b^2$ in two cases where the arms a, b are comparatively small. The standard unit of length is used, of course.

a	b	c (Euclidean)	c (hyperbolic)
0.3	0.4	0.5	0.5046
0.03	0.04	0.05	0.0504

We note how close the values of c given by the hyperbolic formula are to those given by the Euclidean formula. Yet the corresponding triangles are of enormous extent, greater by far than any triangles dealt with by scientists. The agreement between the values of c would be even closer for the triangles encountered in scientific work inasmuch as the values of a and b for those triangles would be very much smaller than those in the table above. In the case of a triangle with arms of 3 and 4 miles, for example, or even 3000 and

4000 miles, the values of a and b would be so small that the value of c given by the hyperbolic formula would actually be indistinguishable from the value given by the Euclidean formula.

Inasmuch as the physical extent of the standard unit of length cannot be precisely determined, might it not be possible to use some other unit of length, one whose physical extent can be determined? In Chapter IV, Section 5, we saw that other units are indeed possible. Their use, however, results in formulas that involve a parameter k and hence are less simple than those we have obtained. Corresponding to $\cosh c = \cosh a \cosh b$, for example, there would be the formula

$$\cosh \frac{c}{k} = \cosh \frac{a}{k} \cosh \frac{b}{k}, \tag{1}$$

corresponding to $\sin \lambda = \sinh a / \sinh c$ there would be

$$\sin \lambda = \frac{\sinh(a/k)}{\sinh(c/k)}, \tag{2}$$

and so forth. The value of k in these formulas depends on the particular unit of length used and a, b, c are the lengths of the sides of the triangle in terms of that unit. If the unit is half of the standard unit, then $k = 2$; if it is a third of the standard unit, then $k = 3$; and so on (IV, §5, Exs. 5 through 7). From this and the discussion earlier in the section we can see that if a unit were used whose physical extent could be determined, then the corresponding value of k would not only be extremely great but actually indeterminate because of the indeterminacy of the physical extent of the standard unit. If a unit were used, for example, which corresponded to a mile when interpreted physically and we wished to find the hypotenuse of a right triangle with arms of 3 and 4 miles, we would substitute 3 and 4 for a and b in formula (1), but, not knowing k, we could not compute c.

Finally, let us return to the formulas employing the standard unit of length and consider the case in which they give results significantly different from those given by Euclidean formulas. The dimensions (in standard units) of a geometric figure need not be very great in order for this to happen. A right triangle, for example, in which $a = 3$ and $b = 4$ has the hyperbolic value $c = 6.3$, whereas the corresponding Euclidean value, of course, is $c = 5$. Interpreted physically, this triangle would be of enormous size, so great, in fact, that astronomers cannot measure it. We therefore do not know which of the two computed values of c is the more accurate length of the hypotenuse. A like uncertainty occurs in all cases where hyperbolic and Euclidean formulas give significantly different results.*

---

* According to relativity theory, hyperbolic formulas may be more accurate in some parts of the universe, and Euclidean formulas more accurate in other parts.

EXERCISES

1. Answer the following questions as completely as possible. (a) If the triangle in Fig. VI, 13 is viewed hyperbolically, what can be said about its angle-sum? (b) Hence what can be said about the angle-sum of any triangle, viewed hyperbolically, which occurs in a familiar engineering project such as the construction of a bridge, tunnel, or tall building?

2. Using the facts given in connection with Fig. VI, 13, what can be said about the size of the angle of parallelism corresponding to any distance occurring in a familiar work of engineering, such as the construction of a road, tunnel, bridge, and so on?

3. In Section 8, Exercise 4(a), the circumference of a circle with a radius of 0.25 was computed and found to be 0.506π. The corresponding Euclidean value is 0.50π. (a) Is the circle small or large when interpreted physically? Justify your answer. (b) Could closer agreement with the corresponding Euclidean value be expected for a circle with a smaller radius? Why?

4. (a) Calculate by formula (1), with $k = 100$, the hypotenuse of the right triangle in which $a = 3$, $b = 4$. (b) Describe the unit segment used in part (a) in relation to the standard unit. (c) Is the triangle in part (a) small or large when interpreted physically? Justify your answer.

5. When is a physical region sufficiently small in order that hyperbolic formulas, applied to figures in the region, should give results in close accord with experience?

6. If the arms and hypotenuse of a right triangle have the lengths a', b', c' in terms of a unit of length which is half the standard unit, prove that

$$\cosh \frac{c'}{2} = \cosh \frac{a'}{2} \cosh \frac{b'}{2}$$

without using formula (1).

7. If λ' denotes the angle opposite side a' in the triangle of Exercise 6, prove that

$$\sin \lambda' = \frac{\sinh(a'/2)}{\sinh(c'/2)}$$

without using formula (2).

8. A circle of radius 1.5 has a circumference of approximately 4.2π according to the hyperbolic formula and a circumference of 3π according to the Euclidean formula. Which value agrees better with physical experience? Justify your answer.

ELLIPTIC GEOMETRY

VII

Double Elliptic Geometry

1. INTRODUCTION

The discovery of non-Euclidean geometry was a very important event in the history of thought, as well as in mathematics. For more than 2000 years prior to the discovery, Euclidean geometry had reigned supreme. Belief in a finite universe had given way to belief in an infinite one, but there was no doubt that its measurements conformed to the canons of Euclidean geometry. All engineering and scientific work had presumed a Euclidean framework for space. A noted philosopher of the eighteenth century, Kant, went so far as to assert that humans were incapable of thinking of physical space in other than a Euclidean way. Although the discovery of non-Euclidean geometry did not make an immediate impact on the intellectual world, it became clear in time that the Euclidean view of the cosmos was only a convenient habit of thought and that one could with equal correctness view the universe as infinite but non-Euclidean.

This was quite revolutionary, to be sure, but the force of the new ideas was too powerful to stop there. Having toppled Euclidean geometry from its pedestal, mathematicians were no longer sure that they fully understood the nature of physical space, or the nature of geometry, or the relation of the two. Accordingly, these age-old problems were subjected to new and thorough scrutiny, with results that were profound for mathematics, science, and philosophy. Other metric geometries besides Euclidean and hyperbolic geometries were discovered, and some of them applied equally well to physical space. The possibility that the universe was finite came to be considered, and consistent with this concept systems of geometry were developed in which the straight line is of finite length. Geometries of more than three dimensions were studied, at first abstractly, but later in connection with the view, as in Einstein's theory of relativity, that the physical universe is best understood not as a three-dimensional entity, but as a four-dimensional one in which time is regarded as having a dimension, along with the three dimensions of space.

2. RIEMANN

A prime mover in this ferment of ideas was the German mathematician G. F. B. Riemann (1826–1866), born at about the time of the discovery of non-Euclidean geometry and later a student of one of its discoverers, Gauss. Mainly as a result of Riemann's thinking, attention was directed toward two new non-Euclidean geometries in which straight lines are finite and in which there are no parallel straight lines. These so-called *elliptic geometries* are the subject of the present chapter and the following one.

Among Riemann's many ideas are two which lead directly to the new geometries. Our experience with physical space strongly suggests that it has no end or boundary, and so, because of its enormity, we customarily regard it as being of infinite extent. Euclidean geometry and hyperbolic geometry are mathematical creations consistent with this concept of space. In the absence of positive scientific confirmation of this view, however, one has an equal right to regard physical space as being of finite, although enormous, extent. So thought Riemann, and he went on to suggest a new concept of a straight line consistent with this view of space. An obvious possibility was to regard straight objects of experience such as taut strings and light rays as being small parts of very long straight paths which come to an abrupt end far off in space. However, preferring to retain for a straight line its familiar property of being endless, Riemann suggested the concept of a *closed* path of finite extent. One who adheres to this view may therefore regard a ruling on a piece of paper or any other physical object commonly called straight, as a tiny part of an enormous closed line in space.

From our knowledge of Euclidean and hyperbolic geometry we can begin to see some of the possible features of a system of geometry in which straight lines are finite. In Euclidean and hyperbolic geometry the existence of parallel lines, we recall, is a consequence of the theorem which states that the exterior angle of a triangle exceeds each opposite interior angle, and this theorem, in turn, is a consequence of the assumption that straight lines are infinite. Hence, a system of geometry whose straight lines are finite could very possibly not have any parallel lines at all. Further, let us recall that Saccheri's hypothesis of the obtuse angle led to a contradiction because straight lines were assumed to be infinite, and that before reaching the contradiction Saccheri had correctly proved several novel propositions, most notably that the angle-sum of a triangle exceeds 180° and that an angle inscribed in a semicircle is obtuse. These properties, together with the property that the summit angles of a Saccheri quadrilateral are obtuse, might therefore reasonably be expected (or at least not unexpected) in a system of geometry whose straight lines are finite.

Before describing the second of Riemann's ideas to which we referred, it will be helpful to call attention to a few simple facts concerning the geometry on a surface in three-dimensional Euclidean space. If A, B are any two points on a surface consisting of a single piece, such as a cone, cylinder, sphere, or paraboloid, then they can be joined, of course, by many different paths, or arcs, on the surface. By the surface distance between the points, or, as it is known technically, the *geodesic distance*, is meant the length of the shortest of these paths. There may be more than one shortest path, depending on the position of A and B. Let (AB) denote a shortest path joining A and B, (AC) a shortest path joining A and C, and (BC) a shortest path joining B and C, the three points being distinct. If these paths meet only in A, B, C, then they form a three-sided figure on the surface called a *geodesic triangle*. The angles of this figure, called *geodesic angles*, are the three angles at which (AB), (AC), (BC) meet in pairs. We need not go any further in order for it to be clear that on each surface there will be a geometry dealing with these shortest paths, geodesic distances, geodesic triangles, and other figures, and that these *surface geometries* may differ from one another because of differences in the surfaces.

Riemann's second idea is then simply the following. Corresponding to each surface geometry there is, generally, an abstract system of geometry which, like Euclidean and hyperbolic geometry, deals with straight line segments, angles, triangles, distance, and so forth, and in which the facts concerning these entities are logically the same as the facts concerning shortest paths, geodesic angles, geodesic triangles, geodesic distance, and so on, in the surface geometry. For example, in sufficiently small regions on a circular cylinder there is a unique shortest path joining each two points, the angle-sum of a geodesic triangle is 180°, and the Pythagorean Theorem holds for geodesic right triangles. As these few facts correctly suggest, the abstract system of geometry which corresponds to the local geometry (that is, the geometry in a small region) on a circular cylinder is Euclidean geometry. The same can be said for the local geometry on a circular cone. There is a surface which is called a *pseudosphere* (a part of it appears in Fig. VII, 1) because it has certain properties in common with a sphere, and in small regions on a pseudosphere there is a unique shortest path joining each two points, the angle-sum of a geodesic triangle is less than 180° and increases as the area decreases, and two geodesic triangles are congruent if their angles are equal, respectively. As these properties suggest, hyperbolic geometry is the abstract system of geometry corresponding to the local geometry on a pseudosphere. The discovery of this fact by E. Beltrami in 1868 provided support to the belief that hyperbolic geometry was logically consistent and thus led to wider acceptance of the system. Clearly, according to Riemann's idea, a key to the discovery of new systems of geometry lies in the study of surface geometries.

Fig. VII, 1

3. THE ELLIPTIC GEOMETRIES

By applying Riemann's second idea to so simple and familiar a surface as a sphere one can become aware of the existence of two different systems of plane geometry in which straight lines are finite and in which there are no parallel straight lines. This awareness will not come, however, unless one is already familiar with certain facts about a sphere, and those facts are brought out in the next section. Here we shall merely name the two systems and offer some basis for distinguishing between them.

In one of the systems any two straight lines meet in a single point. This system is called *single elliptic geometry* or simply *elliptic geometry*. As we shall see, the existence of this system can be revealed by considering the relations which occur on just half of a sphere. The system is therefore also known as *elliptic geometry of the hemispherical type*.

In the other system any two straight lines meet in two points. This system is called *double elliptic geometry*. It is also known as *elliptic geometry of the spherical type* because we are made aware of its existence by considering relations involving the entire sphere.

Each of the elliptic geometries can be developed from axioms dealing with points and straight lines, just as in the case of Euclidean and hyperbolic geometry, and these developments offer much that is novel because of the changed character of a straight line and the absence of parallel lines. Such a logical presentation of double elliptic geometry is given later in the chapter after the student has had an opportunity to become acquainted with the system in an informal way. A similar procedure is followed for single elliptic geometry in the next chapter.

4. GEOMETRY ON A SPHERE

We shall now state, without proof, the facts about a sphere which will be useful to us in our work with the elliptic geometries.

The largest circles on a sphere are those whose radii equal the radius of the sphere. They are called *great circles*. Among the great circles on a terrestrial globe, for example, are the meridian circles, or circles of longitude, each passing through the north and south poles. Circles of latitude, except for the equator, are not great circles.

A pair of great circles always meet in two antipodal points, that is, in two points which are the ends of a diameter of the sphere. All the great circles which go through a point meet again in the antipodal point. Hence, through two antipodal points there pass infinitely many great circles. Through two nonantipodal points, however, there passes a unique great circle.

The shortest path, or line, on a sphere joining one point to another is called a *geodesic arc on the sphere* (for brevity we shall merely say *geodesic arc*). A geodesic arc is always an arc of a great circle. Hence a captain who wishes to take his ship along the shortest route joining two places will follow a great circle course whenever he is free to do so. If two points A, B on a sphere are not antipodal, there is a unique geodesic arc connecting them, namely, the shorter of the two arcs into which A and B divide the great circle through them. If A, B are antipodal, they divide every great circle through them into equal arcs each of which is a geodesic arc. Hence infinitely many geodesic arcs join two antipodal points, each arc being half as long as a great circle.

By the *distance* between two points on a sphere we shall mean the length of the geodesic arc joining them if they are nonantipodal, or, if they are antipodal, the common length of all the geodesic arcs joining them. If A, B, C are any three points on a sphere, then $AB + BC \geq AC$, where AB denotes the distance between A and B, and BC, AC have similar meanings. For a sphere of radius r the maximum possible distance is πr, which is the distance between two antipodal points. The length of each great circle on this sphere is $2\pi r$, or twice the maximum distance, and the area of the sphere is $4\pi r^2$. All the great circles which are perpendicular to any given great circle c meet one another in the same two antipodal points. These points, called the *poles* of c, are all at the same distance $\pi r/2$ from the points of c. This distance is known as *polar distance* on the sphere.

If A, B, and C are distinct points not on the same great circle, then each two of them are nonantipodal and have a unique geodesic arc (Ex. 6). The figure consisting of A, B, C and the geodesic arcs AB, AC, BC is called the *spherical triangle ABC*. The points and geodesic arcs are called the *vertices* and *sides* of the spherical triangle. The *angles* of the spherical triangle are defined to be the three angles, each less than 180°, formed by the three sides in pairs. Two spherical triangles are called *congruent* if the sides and angles of one are equal, respectively, to the sides and angles of the other. It can be shown that two spherical triangles are congruent under the same conditions as are two triangles in hyperbolic geometry except in the case of angle–angle–side.*

The angle-sum of a spherical triangle exceeds 180°, or π radians. The amount of this excess is called *spherical excess* and has the value $\lambda + \mu + \nu - \pi$, where λ, μ, ν are the radian measures of the angles of the triangle. The greater the angle-sum of the triangle, the greater its area, the exact relation being $S = r^2(\lambda + \mu + \nu - \pi)$, where S denotes the area and r the radius of the sphere. Thus S is proportional to the spherical excess, with the constant of proportionality r^2. The similarity of this formula to the corresponding formula in hyperbolic geometry should be noted.

If ABC is a spherical triangle with a right angle at C and with its other parts labeled as in hyperbolic geometry (Fig. VII, 2), then it can be shown that

$$\sin \lambda = \frac{\sin(a/r)}{\sin(c/r)}, \qquad \cos \lambda = \frac{\tan(b/r)}{\tan(c/r)}, \qquad \tan \lambda = \frac{\tan(a/r)}{\sin(b/r)},$$

$$\cos \frac{a}{r} \cos \frac{b}{r} = \cos \frac{c}{r},$$

where r is the radius of the sphere. Analogous formulas hold for $\sin \mu$, $\cos \mu$, $\tan \mu$. In all these formulas λ and μ may be in any angular unit, but the values of a/r, b/r, c/r must be in radians. Thus, if $a = 1$, $r = 2$, then

$$\sin \frac{a}{r} = \sin \frac{1}{2} = 0.47943.$$

---

* Two spherical triangles are not necessarily congruent if two angles and the side opposite one are equal, respectively.

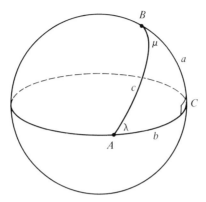

Fig. VII, 2

For the general spherical triangle ABC, with its parts labeled as in hyperbolic geometry, we have the *spherical Law of Sines*

$$\frac{\sin(a/r)}{\sin \lambda} = \frac{\sin(b/r)}{\sin \mu} = \frac{\sin(c/r)}{\sin v}$$

and the *spherical Law of Cosines*

$$\cos \frac{a}{r} = \cos \frac{b}{r} \cos \frac{c}{r} + \sin \frac{b}{r} \sin \frac{c}{r} \cos \lambda,$$

with analogous equations involving μ and v.

All the facts we have stated concern points and figures on a sphere and thus belong to the subject known as *spherical geometry*. The proofs of these facts would be achieved by regarding the sphere as a surface in three-dimensional Euclidean geometry and applying the appropriate theorems of this geometry to the surface. Thus, spherical geometry is simply a branch of Euclidean geometry.

EXERCISES

1. A *circle* in spherical geometry is a locus of points on the sphere at a constant distance from a point on the sphere (the distance being measured on the sphere). The distance and point are called the *radius* and *center* of the circle. If r is the radius of the sphere, verify that (a) a circle with radius $\pi r/2$ is a great circle, (b) a circle with radius πr is a point, (c) a circle with center P and radius a is also a circle with center P' and radius $\pi r - a$, where P' is antipodal to P.

2.   Describe the locus of points on a sphere of radius r whose distance from a given point on the sphere is (a) $\leq \pi r/2$, (b) $\leq \pi r$.

3.   Prove that the circumference C of a circle of radius a in spherical geometry (see Ex. 1) is given by the formula

$$C = 2\pi r \sin \frac{a}{r},$$

where r is the radius of the sphere.

4.   The area S enclosed by a circle of radius a in spherical geometry (see Ex. 1) is given by the formula

$$S = 2\pi r^2 \left(1 - \cos \frac{a}{r}\right),$$

where r is the radius of the sphere. Deduce from this the formula for the area of a sphere of radius r.

5.   Show how to obtain a spherical triangle all of whose angles are right angles. How long are the sides if r is the radius of the sphere?

6.   Verify that three points on a sphere necessarily lie on the same great circle if two of them are antipodal.

7.   Show by an example that an exterior angle of a spherical triangle may equal an opposite interior angle of the triangle.

8.   In Euclidean and hyperbolic geometry if three points A, B, C are encountered in this order when a line is traversed, then $AB + BC = AC$. Show that this distance relation does not always hold when a great circle is traversed.

9.   Given that a point of a geodesic arc divides it into two arcs, prove that they are also geodesic arcs.

10.   Find two spherical triangles which satisfy the condition of angle–angle–side but are not congruent.

5.   A DESCRIPTION OF DOUBLE ELLIPTIC GEOMETRY

The concept of a two-dimensional system of geometry which, like Euclidean and hyperbolic geometry, employs the terms point, straight line, plane, straight line segment, distance, angle, congruent, perpendicular, and so forth, and whose points, straight lines, and straight line segments have the same properties and relations as the points, great circles, and geodesic arcs on a sphere, is the concept of double elliptic plane geometry. Determining the facts of this system is therefore a comparatively simple matter,* requiring mainly a change of wording in the statements of spherical geometry and the replacement of r in its formulas by some other letter, usually k. The term "great circle," for example, is changed to "straight line," "sphere" is changed to "double elliptic plane," "geodesic arc" to "straight line segment," "spherical triangle" to "rectilinear triangle" or simply "triangle," and so forth. We list below a substantial number of facts of double elliptic geometry obtained in this way. No attempt has been made to classify them as axioms, definitions, and theorems. To aid the student in determining the facts of spherical geometry from which they were obtained, we preface the list with a "dictionary" of corresponding terms.

Corresponding Terms

Spherical Geometry	Double Elliptic Geometry
great circle	straight line
geodesic arc	straight line segment
spherical triangle	triangle
sphere	double elliptic plane
antipodal points	opposite points
r	k
radian	standard angular unit

Some Basic Facts of Double Elliptic Geometry

1.   Each pair of straight lines meet in two points.

2.   Through each pair of points there passes at least one straight line.

3.   Through each point there pass infinitely many straight lines, the totality of whose points constitutes the double elliptic plane.

4.   Associated with any two points A, B is a positive number called the *distance* between the points. It is often denoted by AB or BA. If A, B, C are any three points, then $AB + BC \geq AC$.

5.   The distance between any two points A, B equals the length of a part of a straight line through them. This part of a straight line is called a *straight line segment AB*, or, briefly, a *segment AB*. There may be more than one segment AB (in fact, there may be infinitely many). The segments (or segment) joining two points are shorter than any other paths joining them.

6.   Any segment may be chosen to serve as unit segment, or unit of length. Once this choice is made, distances in the double elliptic plane may attain, but not exceed, a certain number of such units. This number is called the *maximum distance*. Two points at the maximum distance are called *opposite points*, each being regarded as opposite the other. There is a unique

---

* To prove these facts rigorously on the basis of axioms, however, is a problem of considerable difficulty.

point opposite any given point. Distinct points have distinct opposite points.

7.   Infinitely many straight lines pass through two opposite points and infinitely many segments connect them. All the straight lines which go through a point also go through the opposite point. A unique straight line goes through each two nonopposite points and a unique segment joins them.

8.   A pair of opposite points divide each straight line through them into two equal parts, which are segments. Conversely, if a pair of points divide a straight line into equal parts, the points are opposite. All straight lines have a length equal to twice the maximum distance.

9.   If the maximum distance is denoted by d, then the length of a straight line is $2d$ and the area of the double elliptic plane is $4d^2/\pi$. Instead of expressing these quantities in terms of d, it is customary to express them in terms of k, representing a certain part of the maximum distance. To be precise, if we let $k = d/\pi$, then the maximum distance is πk, the length of a straight line is $2\pi k$, and the area of the double elliptic plane is $4\pi k^2$. (These formulas are seen to differ from the corresponding formulas of spherical geometry only in that k replaces r.) If we imagine a straight line divided into 2π equal parts, then k is the length of each part.

10.   All the straight lines perpendicular to any given straight line meet in a pair of opposite points. These points, called the *poles* of the line, are at the distance $\pi k/2$, that is, half of the maximum distance, from each point of the line. Also, every point in the plane at the distance $\pi k/2$ from each pole is on the line.

11.   If three points are noncollinear, then each two of them are nonopposite and hence can be joined by only one segment. A *triangle* is the figure consisting of three noncollinear points and the three segments determined by them. The definition of the congruence of two triangles and the conditions for their congruence are the same as in hyperbolic geometry except for the case of two angles and the side opposite one of them.

12.   The angle-sum of a triangle exceeds 180° or π standard units. The area S of a triangle is given by the formula $S = k^2(\lambda + \mu + \nu - \pi)$, where λ, μ, ν are the measures of the angles in standard units.

13.   If ABC is a triangle with a right angle at C and the other parts are labeled in the usual way, then

$$\sin \lambda = \frac{\sin(a/k)}{\sin(c/k)}, \qquad \cos \lambda = \frac{\tan(b/k)}{\tan(c/k)}, \qquad \tan \lambda = \frac{\tan(a/k)}{\sin(b/k)},$$

$$\cos \frac{a}{k} \cos \frac{b}{k} = \cos \frac{c}{k},$$

where the values a/k, b/k, c/k are to be interpreted as numbers of standard angular units. Analogous formulas hold for $\sin \mu$, $\cos \mu$, $\tan \mu$.

14.   For the general triangle ABC labeled in the usual way we have the *double elliptic Law of Sines*

$$\frac{\sin(a/k)}{\sin \lambda} = \frac{\sin(b/k)}{\sin \mu} = \frac{\sin(c/k)}{\sin \nu}$$

and the *double elliptic Law of Cosines*

$$\cos \frac{a}{k} = \cos \frac{b}{k} \cos \frac{c}{k} + \sin \frac{b}{k} \sin \frac{c}{k} \cos \lambda,$$

with analogous equations involving $\cos \mu$ and $\cos \nu$, where a/k, b/k, c/k are interpreted as in item 13.

All the double elliptic formulas we have given differ from the corresponding formulas of spherical geometry only in that k replaces r. If a sphere of radius r is given and a unit of length is chosen in the double elliptic plane so that $k = r$, then the two sets of formulas are identical. For example, if a sphere is given with $r = 1/\pi$ and we choose the longest segment in the double elliptic plane to be the unit of length, then the maximum distance πk is 1, $k = 1/\pi$, and the two sets of formulas are identical.

EXERCISES

Except when otherwise indicated, each exercise refers to double elliptic geometry.

1.   Give the statements of spherical geometry corresponding to the following statements of double elliptic geometry listed in this section:

(a) 1, 2, 3;   (b) 4, 5, 6;   (c) 7, 8, 9;   (d) 10, 11, 12;   (e) 13, 14.

2.   State the facts of double elliptic geometry corresponding to the following facts of spherical geometry:

(a)   The shortest path (or line) joining two points on a sphere is a geodesic arc.
(b)   All the great circles on a sphere have the same length.
(c)   The sum of the lengths of two sides of a spherical triangle exceeds the length of the third side.
(d)   There is a unique point on a sphere which is at the maximum distance from a given point.
(e)   Through each point of a great circle there passes a unique great circle perpendicular to the given circle.
(f)   An exterior angle of a spherical triangle does not always exceed each opposite interior angle.

3. (a) Define a circle in double elliptic geometry. (b) What is the nature of a circle of radius $\pi k/2$? of radius πk?

4. What is the nature of the set of points in double elliptic geometry whose distance from a given point is less than or equal to πk?

5. The formula for the circumference of a circle of radius a in double elliptic geometry is

$$C = 2\pi k \sin \frac{a}{k}.$$

Deduce from it the formula for the length of a straight line [see Ex. 3(b)].

6. Show that circles of radii a and $\pi k - a$ have equal circumferences (see Ex. 5).

7. The formula for the area S of a circle of radius a in double elliptic geometry is

$$S = 2\pi k^2 \left(1 - \cos \frac{a}{k}\right).$$

Show that S approaches the area of the double elliptic plane when a approaches πk.

8. Circles of radii a and $\pi k - a$ have equal circumferences according to Exercise 6. Do they have equal areas (see Ex. 7)? If not, what is the relation between their areas?

9. Given that $\sin x < x$ when $0 < x < \pi$, prove that the circumference of a circle (other than a point-circle) is less than 2π times the radius (see Ex. 5).

10. Using the fact that $\cos x > 1 - (x^2/2)$ when $0 < x < \pi$, show that the area of a circle (other than a point-circle) is less than π times the square of the radius (see Ex. 7).

11. Show that the angle-sum of a triangle decreases as the area decreases and, in particular, approaches π when the area approaches 0.

12. Prove that the double elliptic formula for the circumference of a circle (see Ex. 5) becomes more nearly Euclidean as the radius approaches zero.

13. Prove that the double elliptic formula for the area of a circle (see Ex. 7) becomes more nearly Euclidean as the radius approaches zero.

14. If A, B, C are three noncollinear points, prove that no two of them can be opposite.

15. If A, B, C are three points such that $AB + BC = AC$, then they are collinear. Prove this.

16. State the converse of the theorem in Exercise 14. Is it true? Justify your answer by making use of the relation between double elliptic geometry and spherical geometry.

17. How does k change (if at all) when the unit of length is increased?

6. DOUBLE ELLIPTIC GEOMETRY AND THE PHYSICAL WORLD

From our work earlier in the chapter we know that the geometry on a physical sphere, such as the earth or moon, is essentially double elliptic if the terms straight line, line segment, triangle, and so on, are used rather than great circle, geodesic arc, spherical triangle, and so on. The geometry on a flat physical surface, too, can be regarded as double elliptic if we imagine the surface to be a very small part of a large sphere and the straight paths on the surface to be geodesic arcs on the sphere. This view simply reverses the familiar one of regarding small portions of the earth's surface as flat. Actually, a flat physical surface such as an engineer's drawing board or a table top is not perfectly flat, but has a small overall curvature. Since a sphere of sufficiently large radius will have this same curvature, the flat surface can be regarded as being, at least approximately, a small part of this sphere. Toward the end of the preceding chapter we saw that flat physical surfaces, although customarily thought of as physical models of parts of Euclidean planes, can with equal correctness be viewed as physical models of parts of hyperbolic planes, and hence that the geometry of the straight paths and other figures on those surfaces can be regarded as hyperbolic as well as Euclidean. What we are now suggesting is that the geometry on these same surfaces can also be viewed as double elliptic.

A simple example may help to clarify these ideas further. Suppose that with a compass and straightedge we draw two figures on a blackboard, one of them a triangle, the other consisting of two lines perpendicular to a third. If we imagine the flat surface of the blackboard to be a very small part of an ideal flat surface of infinite extent, then we will think of the angle-sum of the triangle as being 180° under the Euclidean view, very slightly less under the hyperbolic view, and of the two perpendiculars as being parts of infinitely long, nonintersecting lines under both views. But if, with equal right, we imagine the surface of the blackboard to be a very small part of an enormous sphere, then we will regard the angle-sum as very slightly more than 180° (this will be justified presently) and the perpendiculars as parts of very long, but finite, closed lines which meet in two distant points. Actual measurement of the triangle and the perpendiculars will not reveal which of the three views is the correct one. Repeated careful measurement of the triangle, for example, will yield varying angle-sums quite close to 180°, some greater, some less, very few (if any) exactly 180°. All our statements would still hold if, instead of the blackboard, we had used any other flat surface, even one of very considerable extent. Scientists do not know whether the physical universe is finite or infinite, but the double elliptic view is consistent with either situation. This cannot be said for the Euclidean and hyperbolic views, with their infinite straight lines.

To justify our statement that the angle-sum of the triangle is very slightly more than 180° under the double elliptic view, consider the double elliptic formula $S = k^2(\lambda + \mu + \nu - \pi)$, which relates the area S and the angles λ, μ, ν of a triangle in standard units. (The formula is given in item 12 of Section 5.) To apply this formula to the blackboard triangle in order to calculate its angle-sum one should take k to equal the radius of the sphere on which the blackboard is imagined to lie and S to equal the area of the triangle. However, for our present purpose we need not know the precise values of these quantities but only that k is very great and S is very small. Then S/k^2, or $\lambda + \mu + \nu - \pi$, is extremely small and the angle-sum $\lambda + \mu + \nu$ is very slightly more than π standard units, or 180°.

EXERCISE

1. If a circle of radius a is carefully drawn on a blackboard, show that under the double elliptic view its circumference is very slightly less than $2\pi a$ (see §5, Ex. 5).

7. AN AXIOMATIC PRESENTATION OF DOUBLE ELLIPTIC GEOMETRY

We turn now to the question of arranging the facts of double elliptic geometry into a deductive system, that is, a system based on axioms. It will be recalled that the problem of choosing axioms for hyperbolic geometry involved nothing more than replacing Euclid's parallel axiom by the contrary statement or some logical equivalent of it, and that, as a result, we were able at once to take Euclid's Propositions 1 through 28 as the initial theorems in our development of hyperbolic geometry. No such simple procedure is available in the case of double elliptic geometry, for the axioms used in proving many of those 28 propositions involve assumptions that do not hold in double elliptic geometry, for example, the assumption that a unique straight line goes through every two points, or that straight lines are infinite, or that there exists a circle with any given radius. In selecting axioms for double elliptic geometry, then, one must start afresh and determine which of the familiar facts in this geometry can serve as a basis for proving all the others.

It was Euclid's great achievement, as we saw in Chapter I, to solve the comparable problem for the geometry of his day.

While not going so far as to offer a complete set of axioms for double elliptic geometry, we do wish to achieve more in the way of systematic proof than was possible in the exercises of Section 5. Accordingly, we have selected a substantial number of facts of double elliptic geometry and arranged them into a deductive system by accepting some of the facts without proof and using them to prove the remaining ones. As is customary, the statements accepted without proof (other than the definitions) will be called "axioms" and those that are proved will be called "theorems." Although our axioms involve familar facts and concepts, some of the terms used in stating them, such as metric space, simple arc, simple closed curve, may be unfamiliar and will therefore be discussed first.

Simple Arcs and Simple Closed Curves in Metric Spaces

Many important sets of points have the property that a number can be associated with any two members A, B of the set. If this number, which is called the *distance* from A to B and denoted by AB, satisfies the following conditions, the set is called a *metric space*:

(1) $AB = BA$.
(2) $AB = 0$ if $A = B$.
(3) $AB > 0$ if $A \neq B$.
(4) $AB + BC \geq AC$.

The Euclidean and hyperbolic planes are familar examples of metric spaces. It should be clear from the description of double elliptic geometry in the preceding section that the double elliptic plane, too, is a metric space.

A set of points with the properties (1) through (4) will, of course, generally possess further properties. It may be a finite set, for example, or it may be an infinite set. Every two points A, B may have a midpoint (that is, a point C different from A, B such that $AB + BC = AC$ and $AB = BC$), or this may not be true. In other words, there are metric spaces of various kinds. The subject of metric spaces is, in fact, very extensive and we wish to do no more with it than to mention that terms such as curve, continuous curve, closed curve, simple closed curve, arc, simple arc, and so forth, which occur in Euclidean geometry, can also be defined for the general metric space, and, in particular, we wish to clarify the terms simple arc and simple closed curve.

What are usually called arcs in elementary geometry, such as arcs of circles or parabolas, are known more precisely as *simple arcs* since they are arcs which do not intersect themselves. The arc shown in Fig. VII, 3, for example, intersects itself and is therefore not a simple arc. Similarly, in the

Fig. VII, 3

general metric space a simple arc *AB*, viewed intuitively, is a continuous or unbroken curve which has *A*, *B* as endpoints and does not intersect itself.*

Circles, ellipses, triangles, quadrilaterals, and other familar types of polygons are examples of *simple closed curves* in Euclidean geometry. Each of these figures, like a simple arc, is a continuous or unbroken line which does not intersect itself, but, unlike a simple arc, each can be traversed completely by starting anywhere on it and proceeding always in the same direction on it. A closed curve which intersects itself and is therefore not a simple closed curve appears in Fig. VII, 4. Similarly, in the general metric space a simple closed curve, viewed intuitively, is a continuous or unbroken curve which does not intersect itself and which can be completely traversed by starting anywhere on it and proceeding in either of the two possible directions on it.*

Fig. VII, 4

The points *A*, *B* of a simple arc *AB* are called its *endpoints*. A point of a simple arc not an endpoint is called an *interior point* of the arc. Any interior point *C* of a simple arc divides the arc into two other simple arcs having *C* as their only common point. Any two points *A*, *B* of a simple closed curve divide it into two simple arcs *AB* which have *A*, *B* as their only common points.

---

* For the student familiar with mappings we mention these precise definitions: *A simple arc* is the homeomorph, or topological image, of a Euclidean segment (including its endpoints) and a *simple closed curve* is the homeomorph of a Euclidean circle.

By the use of limits the *length* of a simple arc or simple closed curve can be defined in metric spaces much as it is defined in Euclidean geometry, and it possesses all the familiar properties. Thus, (1) every simple arc or simple closed curve has a length (which is either a positive number or infinite); (2) in any subdivision of a simple arc (or simple closed curve) of finite length into two simple arcs the sum of the lengths of these two arcs equals the length of the given arc (or closed curve); (3) on any simple arc AB of finite length s there is a unique point C such that the simple arc AC has any prescribed positive value less than s; (4) if A is any point on a simple closed curve of finite length, there is a unique point B on the curve such that A, B divide it into equal arcs. We shall take these properties for granted.

Aided by the preceding discussion of simple arcs and simple closed curves in metric spaces, we can now begin our axiomatic development of double elliptic geometry. The entire set of points with which we deal, that is, the double elliptic plane, will be denoted by $\mathscr{D}$. Not all of the axioms will be stated immediately. The presentation is divided into several parts.

Straight Lines and Segments

Axiom 1. $\mathscr{D}$ *is a metric space containing at least two points, at least one simple arc of finite length joining each two points, and at least two straight lines.*

Axiom 2.   *A straight line is a simple closed curve of finite length.*

The effect of Axiom 2 is merely to assume that straight lines belong to a certain very extensive class of curves. The axiom therefore does not completely characterize a straight line and one would be wrong to regard it as a definition. In fact, we shall never define a straight line. Whatever properties it has are contained in the axioms and theorems. Similarly, we shall never define a point. This is in keeping with the fact that in every deductive system, of which Euclidean geometry, hyperbolic geometry, and double elliptic geometry are examples, there must be some *undefined terms* as well as some unproved assertions (the axioms).*

Since simple arcs and simple closed curves are infinite sets of points, we have the following immediate consequence of Axioms 1 and 2:

Theorem 1. $\mathscr{D}$ *contains infinitely many points and not all of them are on the same straight line.*

* To define all terms or to prove all assertions would involve one in circularities.

Axiom 3. *Each pair of straight lines meet in exactly two points.*

Axiom 4. *Among the simple arcs joining each two points there is one (or more) whose length is least, that is, no other simple arc joining the points has a smaller length. The length of this simple arc (or of each of them if there is more than one) equals the distance between its endpoints.*

Definition 1. A simple arc with the property stated in Axiom 4 is called a *segment*. If there is exactly one segment joining two points, we shall refer to it as *the segment* joining the points and call it *unique*. $\overline{AB}$ will denote a segment with endpoints A, B.

Axiom 5. *At least one of the simple arcs into which a straight line is divided by each two of its points is a segment.*

From Axiom 5 and Definition 1 we deduce

Theorem 2. *If the two simple arcs into which a straight line is divided by a pair of its points are unequal (in length), only the shorter is a segment joining the points; if they are equal, both are segments joining the points.*

Definition 2. Two points which divide a straight line into equal arcs are called *opposite points on the line*, and each point is said to be *opposite the other on the line*.

It is useful to combine Definition 2 and Theorem 2 as follows.

Theorem 3. *The equal arcs into which each two opposite points on a straight line divide the line are segments joining the points.*

Axiom 6. *Each segment joining two points is (a simple arc) on some straight line through the points.*

If A, B are any two points, there is at least one segment joining them according to Axiom 4 and Definition 1, and there is a straight line through them which contains this segment by Axiom 6. Thus we have

Theorem 4. *There is at least one straight line through each two points.*

Axiom 7. *A straight line through two points is the only straight line through them if they are nonopposite points on the line.*

Theorem 5.   *A straight line g through two points is not the only straight line through them if they are opposite points on g.*

Proof.   Let A, B be opposite points on g. Since, by Theorem 1, not all points of $\mathscr{D}$ are on g, let C be a point not on g. By Theorem 4 there is a straight line through A and C. Denote it by h. Lines g and h, which are distinct, have one common point other than A (Ax. 3)—call it X. Assume $X \neq B$. Then X is an interior point on one of the two segments into which A and B divide g. It follows that the two arcs into which A and X divide g are unequal. Therefore, by Definition 2, A and X are nonopposite points on g. According to Axiom 7, then, g is the only straight line through A and X. But this contradicts that h also goes through them. It follows that $X = B$. Thus h, as well as g, goes through A and B.

Theorem 6.   *A unique straight line goes through two points if and only if there is a unique segment joining them.*

Proof.   Let A, B be two points. Assume there is a unique segment m joining them. By Axiom 6 there is a straight line through A, B which contains m. Suppose h is a second straight line through A, B. Then h meets g only in A, B by Axiom 3, and hence cannot contain m. By Axiom 5 at least one of the simple arcs into which h is divided by A, B is a segment joining A and B. Since this segment must be different from m, we have reached a contradiction of our assumption that there is a unique segment joining A, B. Thus g is the only straight line through A, B.

Now we shall assume that there is more than one segment AB and show that there is more than one straight line AB. Let m, n be two segments AB. Then m, n have equal lengths (Ax. 4, Def. 1). By Axiom 6, m is on some straight line through A, B, and likewise for n. If these lines are distinct, then our proof is finished. If they are the same line, however, then the line, being a simple closed curve, consists of nothing but m and n, which have only A and B in common. Since m and n have equal lengths, A and B are opposite points on the line (Def. 2). By Theorem 5 the line is not the only line through A and B. This completes the proof.

It is easily seen that Axiom 7 implies the following:

Theorem 7.   *Two points cannot be opposite on one straight line and nonopposite on another.*

Next, let us suppose that A is any point, g is a straight line through A, and B is the point on g opposite A. By Theorem 5, g is not the only straight line through *A* and *B*. In fact, examination of the proof of Theorem 5 will show that every straight line through *A* must also go through *B*, and vice versa. In other words, the straight lines through *A* are identical with those through *B*. By Theorem 7, *A* and *B* are opposite points on each of these lines. Thus we have proved

Theorem 8.   *All the straight lines through any given point A also pass through a second point B. The point B is opposite A on each of these lines.*

In view of Theorem 8, if *B* is opposite *A* on one straight line through *A*, it is opposite *A* on every straight line through *A* and we may simply say that *B* *is opposite A* without specifying any particular straight line. Similarly, we shall henceforth say that two points are nonopposite without specifying the straight line on which they are nonopposite. The truth of the following theorem is then apparent.

Theorem 9.   *Corresponding to each point in $\mathcal{D}$ there is a unique opposite point. Any straight line through one of two opposite points also goes through the other. The points of intersection of any two straight lines are opposite. A unique straight line goes through two nonopposite points.*

Now let *g*, *h* be any two straight lines. The two points in which they meet are opposite on both *g* and *h* by Theorem 8 and hence divide each of these lines into equal arcs. These equal arcs on *g* are segments by Theorem 2, and likewise for the equal arcs on *h*. We thus have four segments joining the two points and, by definition of a segment (Def. 1 and Ax. 4), they must be equal. Consequently *g* and *h* must have equal lengths. This proves

Theorem 10.   *All the segments resulting from the intersections of straight lines in pairs are equal. Hence all straight lines have the same length.*

According to this theorem, if *g*, *h* are any two straight lines, not only are the four segments resulting from their intersection equal to each other but they are also equal to the segments resulting from the intersection of any other pair of straight lines. The common value of the lengths of these segments depends, of course, on the choice of the unit segment, and hence can be anything we wish to make it. As explained in Section 5, it is customary to take this common value to be πk, where *k* depends on the choice of the unit segment. We formalize this as follows.

Definition 3.   The segments formed when two straight lines meet are called *half-lines*. The term *half-line ACB* will mean a half-line with endpoints *A*, *B* and interior point *C*. (The endpoints of a half-line are always opposite by Theorem 9.) We take πk as the common length of all half-lines and $2\pi k$ as the common length of all straight lines. The value of k depends on the choice of the unit segment, but in all formulas it is to be regarded as the same unknown constant.

Theorem 11.   *The distance between two points never exceeds πk. It is πk for opposite points and less than πk for nonopposite points.*

Proof.   Let A, B be any two points. By Axiom 4 there is a segment m joining them, its length being equal to the distance AB, and by Axiom 6 it is contained in some straight line g through A, B. If A, B are opposite, they divide g into equal parts which are segments by Theorem 2 and m is necessarily one of them. Being a half-line, m has the length πk (Def. 3), which therefore equals the distance AB (Ax. 4). If A, B are nonopposite, they divide g into unequal parts and m is the shorter part (Theo. 2). The length of m, being less than the length of a half-line, is less than πk, and hence so is the distance AB (Ax. 4)

The truth of the following theorem is easily verified.

Theorem 12.   *There is a unique segment joining two points, and hence a unique straight line through them, if and only if the distance between them is less than πk.*

Theorem 13.   *All the points of $\mathscr{D}$, the double elliptic plane, lie on the straight lines through any two opposite points.*

Proof.   Let A, B be any two opposite points and let P be any point of $\mathscr{D}$. If P is different from A and B, there is a straight line through A and P (Theo. 4), and it goes through B by Theorem 8. Thus P lies on that line.

The following is a useful corollary of Theorem 13.

Theorem 14.   *If three points are noncollinear,* then no two of them are opposite.*

Bilaterals, Angles, and Triangles

Definition 4.   The figure formed by two noncollinear half-lines with the same endpoints is called a *bilateral*. The endpoints and the half-lines are called the *vertices* and the *sides* of the bilateral. Figure VII, 5 shows a bilateral with sides ACB, ADB and vertices A, B.

---

* As usual, *noncollinear* will mean " not on a straight line."

Theorem 15.   *Each two points of a bilateral other than the vertices are nonopposite.*

Proof.   Consider a bilateral *t* with vertices *A, B* (Fig. VII, 5). Let *C* be a third point of *t*. The points *A, C* are clearly nonopposite, and likewise *B, C*. Let *D* be a fourth point of *t*. If *C, D* are on the same side of *t*, they are clearly nonopposite. If they are on different sides, as in Fig. VII, 5, assume they are opposite. There is a straight line containing half-line *ACB* (Ax. 6) and it is

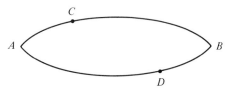

Fig. VII, 5

identical with straight line *AC* (Ax. 7). Straight line *AC* therefore contains half-line *ACB*. Also, since *C, D* are assumed to be opposite, straight line *AC* goes through *D* inasmuch as it goes through *C* (Theo. 8). Similar reasoning shows that straight line *AD* contains half-line *ADB* and goes through *C* inasmuch as it goes through *D*. Straight lines *AC, AD* are identical by Axiom 7 since they both contain the nonopposite points *A, C*. Half-lines *ACB, ADB* are therefore collinear. This contradicts that they form a bilateral. Hence *C, D* cannot be opposite.

EXERCISES

1.   Is the converse of Theorem 14 true? Justify your answer.

2.   A point *C* is said to be *between* two others *A, B* if it is different from them and *AC + CB = AB*. Describe the locus of points between (a) two nonopposite points, (b) two opposite points.

3.   If *A, B, C* are three collinear points, show by an example that no one of them is necessarily between the other two (see Ex. 2).

4.   If *C* is between *A* and *B* (see Ex. 2), and *B* is between *C* and *D*, is *C* necessarily between *A* and *D*? Justify your answer.

Definition 5.   If *A, B* are opposite points, the half-line *ACB* exclusive of *B* is called the *ray AC* and exclusive of *A* is called the *ray BC*. The *endpoint* of ray *AC* is *A*; the endpoint of ray *BC* is *B*. The figure consisting of two rays with a common endpoint is called an *angle*; the endpoint and rays are called the *vertex* and *sides* of the angle. If the sides of an angle are collinear, the angle is called a *straight angle*. Any angle not a straight angle can be regarded as the figure that results when one vertex is removed from a bilateral. An angle with vertex A and sides AC, AD is denoted by $\angle CAD$ or $\angle DAC$. Figure VII, 5 shows a bilateral with vertices A, B and the angles $\angle CAD$, $\angle CBD$ which result from excluding B and A, respectively, from the bilateral.

Two points on different sides of an angle other than a straight angle are nonopposite (Theo. 15) and have a unique segment. Each interior point of this segment is called an *interior point of the angle*. A ray whose endpoint coincides with the vertex of an angle (not a straight angle) and which contains an interior point of the angle is called a *subdivider of the angle*. The straight line containing the ray will also be called a subdivider of the angle. A *subdivider of a straight angle* is any ray (not a side of the angle) whose endpoint coincides with the vertex of the angle.

Definition 6. If A, B are the vertices of a bilateral (Fig. VII, 5) and C, D are two other points of the bilateral not on the same side, we call $\angle CAD$, $\angle CBD$ the *angles of the bilateral*. These angles are never straight angles since the sides of a bilateral are noncollinear.

Angles can be compared in size just as in Euclidean geometry and the logical details of this procedure will therefore not concern us. The *measure*, or value, of an angle is always a positive number, not exceeding 180 when degrees are used, or π when standard units are used. The terms right angle, acute angle, obtuse angle, complementary angles, supplementary angles, vertical angles, perpendicular, and so on, have their usual meaning. A right angle, for example, is greater than an acute angle and less than an obtuse angle; the sum of the measures of complementary angles is $90°$; the sum of the measures of supplementary angles is $180°$; and so forth.

In addition to these familiar terms concerning angles there are also a number of familiar properties of angles that hold in double elliptic geometry. Although some of them can be deduced from others, it will be simpler for our purpose to assume them all without proof. This is done in Axiom 8, where the properties are referred to collectively as the *angular theory at a point*. By this we mean the following:

1. At each point of intersection of two straight lines there are formed two pairs of vertical angles, the angles in each pair being equal.

2. A straight line through a common point of two other straight lines subdivides just one pair of vertical angles formed by the two lines at the point.

3. All straight angles are equal, with the common value $180°$.

4.   All right angles are equal, with the common value 90°.

5.   At each point of a straight line there is a unique perpendicular to the line.

6.   Every angle can be bisected, and uniquely.

7.   If $\angle ABC$ is any angle (Fig. VII, 6) and ray BD subdivides it, then $\angle ABC = \angle ABD + \angle CBD$.

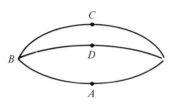

Fig. VII, 6

8.   If $\angle ABC$ is any angle, there is a unique ray BD which subdivides it so that $\angle ABD$ and $\angle CBD$ have specified values. (Properties 5 and 6 are special cases of this.)

Axiom 8.   *The angular theory at a point is the same as in Euclidean geometry.*

Triangles

Definition 7.   The figure consisting of three noncollinear points A, B, C and the three segments $\overline{AB}$, $\overline{AC}$, $\overline{BC}$ is called the *triangle ABC*. The points and segments are called the *vertices* and *sides* of the triangle. The *angles of the triangle* are the angles $\angle ABC$, $\angle BCA$, $\angle CAB$ (they may be abbreviated $\angle B$, $\angle C$, $\angle A$). Two triangles are called *congruent* if the sides and angles of one are equal, respectively, to the sides and angles of the other.

Clearly, no two sides of a triangle are collinear. Hence each angle is less than 180°. We note, further, that each two vertices are nonopposite (Theo. 14), that there is just one segment joining them, and hence that there is always a unique triangle with given vertices.

In his study of Euclidean geometry, which very likely was intuitive in many ways, the student may recall that proofs were given for all of the propositions on congruent triangles. In more rigorous treatments of the subject, however, this is sometimes found not to be possible and, instead, one of these propositions is assumed without proof, as an axiom. We shall do likewise in developing double elliptic geometry.

Axiom 9. *Two triangles are congruent if two sides and the included angle of one triangle are equal, respectively, to two sides and the included angle of the other.*

Theorem 16. *The angles of a bilateral are equal.*

Proof. Let *A*, *B* be the vertices and *e*, *f* the sides of a bilateral (Fig. VII, 7). Let *C* be the midpoint of *e*, and *g* the straight line through *C* perpendicular to *e*. Then *g* meets the straight line containing *f* in two points and

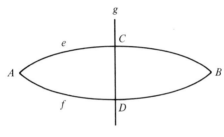

Fig. VII, 7

they are opposite (Theo. 9). Hence (since *f* is a half-line) exactly one of these points is on *f*, and it cannot be *A* or *B*. Denote it by *D*. Since lines *AC*, *CD* are distinct, *D* could be on line *AC* only if it were opposite *C*, but this is impossible (Theo. 15). Hence *A*, *C*, *D* are noncollinear, and the same can be shown for *B*, *C*, *D*. Thus we obtain triangles *ACD*, *BCD*, and they are congruent by Axiom 9 since $AC = BC$, $\angle ACD = \angle BCD$, $CD = CD$. Hence $\angle CAD = \angle CBD$. These are the angles of the bilateral.

By carrying the preceding proof a little further, one can show that the next theorem is also true. We leave this as Exercise 1.

Theorem 17. *The straight line which bisects one side of a bilateral at right angles also bisects the other at right angles.*

The following corollary of Theorem 16 will be useful.

Theorem 18. *Two straight lines which are perpendicular at one of their common points are also perpendicular at the other.*

Analogous to Theorem 15 we have

Theorem 19.   *Each two points of a triangle are nonopposite.*

The proof of this is left as Exercise 2.

Next, let us take any vertex of a triangle, say A, and an interior point D on the side opposite A. Then, by Theorem 19, there is a unique segment joining A and D, and hence a unique straight line. Also, by Definition 5 this line is a subdivider of $\angle A$. Thus we have

Theorem 20.   *A straight line which goes through a vertex of a triangle and an interior point on the opposite side subdivides the angle at that vertex.*

The next axiom consists of two parts. The first of these is the converse of Theorem 20 and the second is just like Pasch's Axiom in Euclidean and hyperbolic geometry.

Axiom 10.   (*a*) *A straight line which subdivides an angle of a triangle intersects the opposite side in an interior point.*

(*b*)   *A straight line which intersects a side of a triangle but does not go through a vertex also intersects one other side.*

In Axiom 10(a) we wished to emphasize that the point of intersection is an interior point, and so included this fact. Had we not included it, it could easily have been proved.

Theorem 21.   *Two triangles are congruent if two angles and the included side of one are equal, respectively, to two angles and the included side of the other.*

Proof.   Let ABC, $A'B'C'$ (Fig. VII, 8) be triangles such that $\angle ABC = \angle A'B'C'$, $AB = A'B'$, $\angle BAC = \angle B'A'C'$. Assume $AC > A'C'$. Take D on $\overline{AC}$ so that $AD = A'C'$. Then A, B, D are noncollinear (Theo. 19) and

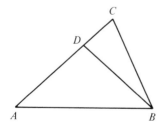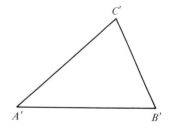

Fig. VII, 8

determine a triangle ABD. This triangle is congruent to triangle $A'B'C'$ (Ax. 9). Hence $\angle ABD = \angle A'B'C'$, and therefore $\angle ABC = \angle ABD$. According to Axiom 8 this is impossible since straight line BD subdivides $\angle ABC$ (Theo. 20). A similar contradiction results from assuming that $A'C' > AC$. Hence $AC = A'C'$. The given triangles are then congruent by Axiom 9.

The proof of the next theorem is left as an exercise.

Theorem 22. *In an isosceles triangle the angles opposite the equal sides are equal and the straight line which bisects the angle formed by those sides bisects the third side at right angles.*

EXERCISES

1. Prove Theorem 17.

2. Prove Theorem 19.

3. Prove Theorem 22.

4. If an angle and an adjacent side of two triangles are equal, respectively, and the other adjacent sides are unequal, then the angles opposite these sides are unequal, the greater angle lying opposite the greater side. Prove this.

5. By an interior point of a triangle ABC with respect to A is meant any interior point of a segment joining A to any point (not B or C) of the opposite side. Show that an interior point of the triangle with respect to A is also an interior point with respect to B and an interior point with respect to C. (Hence we define an *interior point of a triangle* to be any point not belonging to the triangle but situated on a segment joining a vertex of the triangle to a point on the opposite side.)

Structure of the Double Elliptic Plane

Theorem 23. *All the straight lines perpendicular to any given straight line meet in the same two opposite points. The distance from each of these points to any point of the given line is $\pi k/2$.*

Proof. Let p be a straight line, A any point on it, and g the straight line through A perpendicular to p (Fig. VII, 9).* Line g meets p in a further

---

* In some of our diagrams, such as Fig. VII, 9, straight lines are drawn to resemble great circles on a sphere because their relations are most easily exhibited that way.

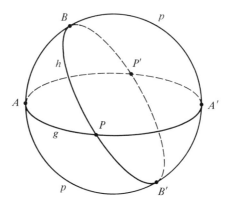

Fig. VII, 9

point A' and is perpendicular to it there (Theo. 18). Let P be the midpoint of one of the half-lines into which the opposite points A, A' divide g. Segments $\overline{PA}$, $\overline{PA'}$ are then unique and

$$PA = PA' = \tfrac{1}{2}\pi k. \tag{1}$$

Let h be any straight line through P other than g. Then h, p meet in a pair of opposite points B, B'. The four nonoverlapping parts into which A, A', B, B' divide p are clearly unique segments, two of them being $\overline{AB}$ and $\overline{A'B'}$. Point P divides one of the half-lines BB' on h into unique segments PB, PB'. The six segments mentioned form two sets of three each, $\overline{PA}$, $\overline{PB}$, $\overline{AB}$ and $\overline{PA'}$, $\overline{PB'}$, $\overline{A'B'}$, which are, respectively, the sides of the triangles PAB and $PA'B'$. Because of (1) and the fact that

$$\measuredangle PAB = \measuredangle PA'B' = 90°, \qquad \measuredangle APB = \measuredangle A'PB', \tag{2}$$

these triangles are congruent (Theo. 21). Hence $PB = PB'$. Since $PB + PB' = \pi k$, we infer that $PB = PB' = \pi k/2$. Using (1) we then obtain $PA = PB = PA' = PB'$. Triangles PAB, $PA'B'$ are thus isosceles. From (2) and Theorem 22 it follows that $\measuredangle ABP$ and $\measuredangle A'B'P$ are right angles. Thus h is perpendicular to p at B and B'. In other words, all straight lines through P are perpendicular to p.

We can now infer, conversely, that all straight lines perpendicular to p must go through P. For let m be a straight line perpendicular to p at Q. Then, by what was proved above, straight line PQ is perpendicular to p at Q. If m were different from straight line PQ, we would have two straight lines perpendicular to p at Q, which is impossible (Ax. 8). Hence m goes through P.

Also, we can infer that the distance from P to each point of p is $\pi k/2$. For if Q is any such point, then the straight line PQ is either g or h of the proof above, Q is one of the points A, A', B, B', and hence $PQ = \pi k/2$.

To complete the proof we note, first, that since all the straight lines perpendicular to p go through P, they must also go through the opposite point P' (Theo. 8). Let us regard g of the first part of the proof as any straight line perpendicular to p, and A as any point on p. Then, since A, A' are opposite on g, and so are P, P', these four points divide g into four nonoverlapping unique segments $\overline{AP}$, $\overline{PA'}$, $\overline{A'P'}$, $\overline{P'A}$ such that, in particular,

$$PA + AP' = \pi k.$$

Combining this with (1) we get $AP' = \pi k/2$. Thus the distance from P' to an arbitrary point A on p is $\pi k/2$.

The opposite points mentioned in Theorem 23 are called the *poles* of the given straight line and their distance $\pi k/2$ from the points of this line is called *polar distance*.

In proving Theorem 23 we also established

Theorem 24. *The set of straight lines perpendicular to any straight line is identical with the set of straight lines through the poles of the line.*

The proof of the following is left to the student.

Theorem 25. *Corresponding to any given point there is a unique straight line for which the point is a pole.*

The straight line mentioned in Theorem 25 is called the *polar* of the point. In other words, if P is a pole of p, then p is the polar of P. Consequently, Theorem 24 can be restated in the following way: *The straight lines through any point are identical with the straight lines perpendicular to the polar of the point.*

How many perpendiculars to a given straight line pass through a given point? If the point is a pole of the line, then of course every straight line through the point is a perpendicular by Theorem 24. If the point is on the line, there is just one perpendicular by Axiom 8, as was seen earlier. The same is true whenever the point is not a pole of the line, as we now show.

Theorem 26. *Through any point not a pole of a given straight line there passes a unique straight line which is perpendicular to the given line.*

Proof. Let A be the point, p the given line, and P a pole of p. Since there is a unique point opposite P, and it is the other pole of p, it cannot be A. Thus A, P are distinct and nonopposite. Hence there is a unique straight line q through them (Theo. 9). This line is perpendicular to p since it goes through P (Theo. 24). Assume r is another straight line through A perpendicular to p. Then r goes through P (Theo. 24). Thus r goes through A and P. This contradicts that q is the only straight line through A, P.

Definition 8. Let A be a point not on a given straight line p, nor a pole of p, and let g be the straight line through A perpendicular to p (Fig. VII, 10). The points B, C in which g meets p are opposite and divide g into two half-lines. Then A is on one of these half-lines, dividing it into two unique segments $\overline{AB}$, $\overline{AC}$ of unequal lengths. The shorter segment is called the *perpendicular from A to p* and its length is called the *distance from A to p*. If, as in Fig. VII, 10, $\overline{AB}$ is the shorter segment, we call B the *foot of the perpendicular from A to p*, or the *projection of A on p*. In Fig. VII, 10, P is a pole of p.

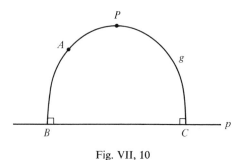

Fig. VII, 10

Theorems 23 and 24 throw much light on the character of the double elliptic plane, that is, the entire set of points $\mathcal{D}$ with which we have been dealing. Consider any straight line p and its poles P, P'. If a variable point starts from P (or P') and proceeds in either direction along any straight line through that point, it will reach p on covering a distance of $\pi k/2$, that is, on traversing a segment of length $\pi k/2$. Thus the nature of the plane is such that the locus of points at the distance $\pi k/2$ from P is identical with the locus of points at the distance $\pi k/2$ from P', the locus being p. Continuing beyond p on the chosen line, the variable point will reach P' (or P) on covering an additional distance of $\pi k/2$. Furthermore, the set of points at a distance less than $\pi k/2$ from P is identical with the set at a distance greater than $\pi k/2$ from P', and the set at a distance greater than $\pi k/2$ from P is identical with the set at a distance less than $\pi k/2$ from P'. In particular, the set of points at the distance πk from either pole consists of a single point, namely, the other pole. The double elliptic plane is, of course, an abstract entity, but all of its above-mentioned properties can be exhibited on a sphere of radius k by interpreting P, P' as antipodal points (Fig. VII, 11), p as the great circle midway between them, and the straight lines through P and P' as the great circles through the two antipodal points.

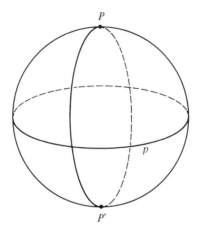

Fig. VII, 11

Retaining the usual definition of a *circle*, namely, that it is a locus of all points at the same distance (called the *radius*) from a point (called the *center*), we can state the following proposition on the basis of the discussion just given.

Theorem 27. *The double elliptic plane consists of any of its points P and the set of circles with center P and radii equal to or less than πk. The circle with center P and radius r is also the circle with center P' and radius $\pi k - r$, where P' is opposite P. In particular, the circle with center P and radius $\pi k/2$ is the straight line whose poles are P, P', and the circle with center P and radius πk consists only of P'.*

Since the straight lines through any point P contain all the points of the double elliptic plane, we see from the discussion accompanying Fig. VII, 11 that an arbitary point of the plane not on p is either at a distance less than $\pi k/2$ from P or at a distance greater than $\pi k/2$ from P. Thus the points of the plane not on p form two separate classes, the points of one class being at a distance less than $\pi k/2$ from P and those of the other class at a distance greater than $\pi k/2$ from P. These classes are called the two *half-planes* determined by p or, briefly, the two *sides* of p, and the double elliptic plane is said to be *two-sided*.* Clearly we have established

Theorem 28.   *The poles of a straight line lie on different sides of the line.*

Other basic facts concerning the sides of a straight line are stated in the next two theorems.

Theorem 29.   *A segment joining two points on different sides of a straight line contains a point of the line.*

Proof.   Let p be the line, P and P' its poles, A and B two points on different sides of p, and $\overline{AB}$ a segment joining A and B. The theorem is easily seen to be true if at least one of the points A, B is a pole of p. Hence we shall suppose that neither point is a pole.

If A, B are collinear with the poles, then $\overline{AB}$ either contains no pole (Fig. VII, 12) or it contains exactly one, say P (Fig. VII, 13). In the first case $\overline{AB}$ is a part of $\overline{AP}$, and in the second case $\overline{AP}$ is a part of $\overline{AB}$. Since the length of $\overline{AP}$ exceeds $\pi k/2$, in both cases there is a point C on $\overline{AP}$ such that $CP = \pi k/2$. Then C is on p. In Fig. VII, 12, C is on $\overline{AB}$ since $BP < \pi k/2$; in Fig. VII, 13, C is on $\overline{AB}$ since $\overline{AP}$ is a part of $\overline{AB}$.

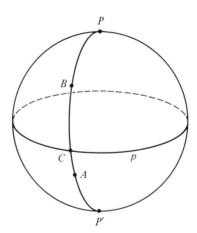

Fig. VII, 12

* The Euclidean and hyperbolic planes are also two-sided. In each case, as we have seen, a straight line divides the plane into two half-planes.

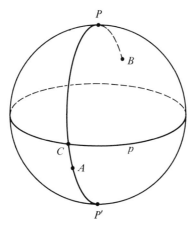

Fig. VII, 13

If A, B are not collinear with the poles (Fig. VII, 14), then A, B, P are the vertices of a triangle. As in the first part of the proof, $\overline{AP}$ meets p in a point C. Since C is an interior point of $\overline{AP}$, p also meets another side of triangle ABP in an interior point D of that side (Ax. 10). This other side cannot be $\overline{BP}$ since $BP < \pi k/2$. It must therefore be $\overline{AB}$. Thus p meets $\overline{AB}$ in D.

In Theorem 29 the path joining the two points is restricted to be a segment. It can be shown that the theorem also holds when the path is any simple arc of a straight line. We leave this as Exercise 2.

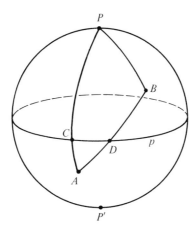

Fig. VII, 14

Theorem 30. (a) *Two opposite points cannot lie on the same side of any straight line.*

(b) *If one endpoint A of a segment is on a given straight line and the other endpoint B is not, the remaining points of the segment lie on the same side of the line as B.*

(c) *If two points are on the same side of a straight line, each point of their segment is on that side.*

(d) *The points of a triangle not on a specified side of the triangle all lie in the same one of the two half-planes determined by the straight line containing that side.*

Proof. (a) Let A, B be opposite points and let p be any straight line not containing them. Assume A, B are on the same side of p. This side contains a pole of p, say P. Hence $AP < \pi k/2$ and $BP < \pi k/2$. Thus $AP + PB < \pi k$. Also, $AB = \pi k$ since A, B are opposite points. It follows that $AP + PB < AB$. But $AP + PB \geq AB$ since the double elliptic plane $\mathscr{D}$ is a metric space (Ax. 1). From this contradiction we see that A, B cannot be on the same side of p.

(b) Let A be on the straight line p and let B be on a side of p. Since the point opposite A is also on p, A and B are nonopposite and have a unique segment AB. Let C be any inner point of $\overline{AB}$. Then C is not on p, for if it were, it would have to be opposite A (Theo. 9), which is impossible since an inner point of a segment cannot be opposite an endpoint. If B, C were on different sides of p, an inner point of $\overline{BC}$, and hence of $\overline{AB}$, would be on p (Theo. 29), which is impossible for the reason given in the preceding sentence. Thus C is on the same side of p as B.

The proofs of (c) and (d) are left as exercises.

EXERCISES

1. Prove Theorem 25.

2. Show that Theorem 29 also holds when the path joining the two points is any simple arc of a straight line.

3. Prove Theorem 30, part (c).

4. Prove Theorem 30, part (d).

5. Show that a point on a straight line and a point not on the line are always nonopposite.

6. If a straight line meets a segment only in an interior point, then the endpoints of the segment are on different sides of the line. Prove this.

7. If A, B, C are points such that A is on a straight line g and B, C are on different sides of g, then g subdivides $\not\prec BAC$. Prove this.

8.   Show that two angles and the side opposite one of them in a triangle can equal two angles and the side opposite one of them in another triangle without the triangles being congruent, unlike the situation in Euclidean and hyperbolic geometry.

Triangle Relations

We next prove several theorems concerning the angles and sides of a triangle.

Theorem 31.   *If a triangle contains a right angle, each remaining angle is acute, right, or obtuse according as the side opposite it is less than, equal to, or greater than $\pi k/2$, and conversely.*

Proof.   Let ABC be a triangle with a right angle at C. We shall prove the theorem for $\angle A$. The proof for $\angle B$ is essentially the same.

(a)   Suppose $BC = \pi k/2$ (Fig. VII, 15). From this and the fact that $\angle C$

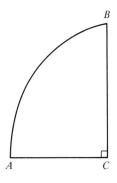

Fig. VII, 15

is a right angle it follows that B is a pole of line AC.* Since the lines through a point are identical with the lines perpendicular to the polar of the point, line AB is perpendicular to line AC. Thus $\angle A$ is a right angle. Conversely, suppose $\angle A$ is a right angle. Then, since lines AB and BC are perpendicular to line AC, their point of intersection B is a pole of line AC, and side AB has a length of $\pi k/2$ (Theo. 23).

---

* From now on "line" always means "straight line."

(b)   Suppose $BC < \pi k/2$. Let P be the nearer pole of line AC to B (Fig. VII, 16). Then A, C, P are the vertices of a triangle in which $\angle CAP$ is a right angle, as well as $\angle ACP$. Line AB, which goes through vertex A of this triangle and interior point B on the opposite side, subdivides $\angle CAP$ (Theo. 20). Hence $\angle BAC$ is acute. Conversely, suppose $\angle BAC$ is acute. Then B is not a pole of line AC, and so $BC \neq \pi k/2$. If we assumed $BC > \pi k/2$, then $\overline{BC}$ would contain a pole P of line AC. Thus line AP would go through vertex A

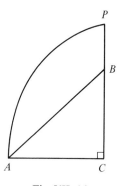

Fig. VII, 16

of triangle ABC and interior point P on the opposite side. It would therefore subdivide $\angle BAC$ (Ax. 10). This is impossible since $\angle BAC$ is acute and $\angle PAC$ is a right angle. It follows that $BC < \pi k/2$.

(c)   Suppose $BC > \pi k/2$. Then, by the converse arguments in (a) and (b), $\angle A$ cannot be a right or acute angle without contradicting the present hypothesis, and hence is obtuse. Conversely, suppose $\angle A$ is obtuse. Then the direct arguments in (a) and (b) show the impossibility of having $BC \leq \pi k/2$. We therefore conclude that $BC > \pi k/2$.

Theorem 32.   *In a right triangle containing two acute angles the hypotenuse exceeds each arm.*

Proof.   Consider triangle ABC, with $\angle C$ a right angle and $\angle A$, $\angle B$ acute (Fig. VII, 17). Then $AB \neq BC$ since $\angle C \neq \angle A$ (Theo. 22). Suppose $AB < BC$. Take D on $\overline{BC}$ so that $BD = AB$. Line AD subdivides $\angle BAC$ (Theo. 20). Hence $\angle BAD$ is acute. Since triangle ABD is isosceles, $\angle BDA$ is also acute (Theo. 22). Hence $\angle CDA$ is obtuse (Ax. 8), so that $AC > \pi k/2$ (Theo. 31). But, since $\angle B$ is acute, $AC < \pi k/2$ (Theo. 31). From this contradiction we infer that $AB > BC$. The proof that $AB > AC$ is similar.

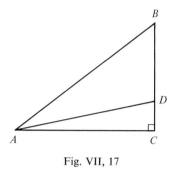

Fig. VII, 17

If the angles at vertices *A* and *B* of a triangle *ABC* are not both right angles, then *C* is not a pole of line *AB* according to Theorem 23. There is, then, a unique line through *C* perpendicular to line *AB* (Theo. 26). In particular, we have

Theorem 33. *If ⊀A and ⊀B of any triangle ABC are acute, the line through C perpendicular to line AB subdivides ⊀C and meets side AB in an interior point.*

Proof. As noted in the paragraph preceding the theorem, there is a unique line through *C* perpendicular to line *AB*. By Theorem 24 it goes through the poles of line *AB*. If *P* is the pole on the same side of line *AB* as *C* (Fig. VII, 18), then the unique line is *PC*. We shall show that it subdivides ⊀*ACB*. By Axiom 8, line *PA* subdivides either ⊀*CAD* or ⊀*CAB*. It cannot subdivide ⊀*CAB* since this angle is acute and ⊀*PAD* is a right angle. It must therefore subdivide ⊀*CAD*. Similarly, line *PB* must subdivide ⊀*CBE*.

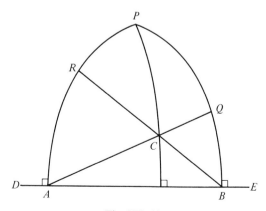

Fig. VII, 18

Hence, also by Axiom 8, lines AC, BC subdivide $\angle PAB$, $\angle PBA$ of triangle PAB. It follows from Axiom 10 that lines AC, BC meet sides PB, PA of this triangle in interior points Q, R. Since lines AC, AR are distinct, C cannot be on line AR unless it is opposite A, which is impossible. Hence A, C, R are non-collinear. Similarly, B, C, Q are noncollinear. Thus we obtain triangles ACR and BCQ. If line PC subdivided $\angle ACR$, it would meet side AR of triangle ACR in a point (Ax. 10) and the point would be opposite P. This is impossible. Similarly, line PC cannot subdivide $\angle BCQ$. It must then subdivide $\angle ACB$ (Ax. 8) and hence meet $\overline{AB}$ in an interior point (Ax. 10).

EXERCISE

1.   The sum of the lengths of any two sides of a triangle exceeds the length of the third side. Show why this is so.

Quadrilaterals

Definition 9.   The figure consisting of four points A, B, C, D, no three collinear (and hence no two opposite), and the unique segments AB, BC, CD, DA is called the *quadrilateral ABCD*. The points and segments are called the *vertices* and *sides* of the quadrilateral, and $\angle ABC$, $\angle BCD$, $\angle CDA$, $\angle DAB$ are the *angles of the quadrilateral*.

Let A, B be any nonopposite points and let P be a pole of line AB (Fig. VII, 19). Take inner points D, C on $\overline{AP}$, $\overline{BP}$ so that $AD = BC$. We call quadrilateral $ABCD$ a *Saccheri quadrilateral*, and $\overline{AB}$ is the *base* of the quadrilateral, $\overline{CD}$ the *summit*, and $\angle C$, $\angle D$ the *summit angles*. It should be

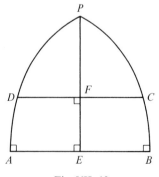

Fig. VII, 19

noted that $\overline{AD}$, $\overline{BC}$ are perpendicular to $\overline{AB}$, that no two sides of the quadrilateral meet except in a vertex, and that all points of the quadrilateral not on $\overline{AB}$ lie on the same side of line AB. All these properties, we recall, are also possessed by Saccheri quadrilaterals in hyperbolic geometry.

Theorem 34.   *The summit angles of a Saccheri quadrilateral are equal and obtuse. The line which bisects the base at right angles does likewise to the summit.*

Proof.   Everything in the theorem except the assertion that the summit angles are obtuse can be proved by the use of congruent triangles, and the details are left as an exercise. Taking for granted that this has been done, we shall go on to prove the specified assertion. Consider the Saccheri quadrilateral $ABCD$ in Fig. VII, 19. Let E, F be the midpoints of $\overline{AB}$, $\overline{CD}$. Then, as is to be shown in Exercise 3, $\overline{EF}$ is perpendicular to $\overline{AB}$ and $\overline{CD}$. Since F is on $\overline{CD}$, it is on the same side of line AB as are C, D [Theo. 30, part (c)], and hence so is each inner point of $\overline{EF}$ [Theo. 30, part (b)]. It follows that since P is the pole of line AB on that side, F is on $\overline{EP}$, points C, F, P are noncollinear, and so are D, F, P. Thus C, F, P are the vertices of a triangle with a right angle at F and with $FP < \pi k/2$. The angle at C in this triangle being acute (Theo. 31), $\angle BCD$ is obtuse. Similarly, using triangle DFP we see that $\angle ADC$ is also obtuse.

Quadrilateral $EBCF$ in Fig. VII, 19 clearly has the following properties (among others): (1) three of its angles are right angles, (2) the two sides included by these angles are each of length less than $\pi k/2$, and (3) the points of the quadrilateral not on a certain one of these sides (namely, $\overline{EB}$) all lie in the same half-plane determined by the line containing that side. We shall call any quadrilateral with these properties a *Lambert quadrilateral*. The angle of the quadrilateral other than the three angles mentioned in property (1) will be called its *fourth angle*.

Lambert quadrilateral $EBCF$ in Fig. VII, 19 was obtained by subdividing Saccheri quadrilateral $ABCD$. Conversely, from any given Lambert quadrilateral a Saccheri quadrilateral can be obtained. We leave the verification of this fact and the proofs of Theorems 35 and 36, which depend on it, to the student.

Theorem 35.   *The fourth angle of a Lambert quadrilateral is obtuse and each side adjacent to it is shorter than the opposite side.*

Theorem 36.   *The summit of a Saccheri quadrilateral is shorter than the base.*

EXERCISES

1. Each two points of a Saccheri or Lambert quadrilateral are non-opposite. Explain why this is so.

2. Show that from any Lambert quadrilateral a Saccheri quadrilateral can be obtained.

3. Prove Theorem 34 except for the statement that the summit angles are obtuse.

4. Prove Theorem 36.

5. Choose any side s of a quadrilateral, including its endpoints. If all the other points of the quadrilateral lie in the same one of the two half-planes determined by the line containing s, then the quadrilateral is said to be *convex*. Show that (a) a Saccheri quadrilateral is convex, (b) a Lambert quadrilateral is convex. [See Theo. 30, part (d).]

6. Taking for granted that two opposite vertices* of a convex quadrilateral (see Ex. 5) lie on different sides of the line through the other vertices, show that the line subdivides the angles at those vertices. (See Ex. 7 preceding Theo. 31.)

7. Prove that the line joining an inner point of a side of a Lambert quadrilateral to a vertex not on that side subdivides the angle at that vertex. (See Exs. 5, 6.)

8. Prove Theorem 35. (The proof of the second part depends on the first part and on Ex. 7.)

Angle-Sum of a Triangle

If a vertex of a triangle is a pole of the line through the other vertices, then the angles at these other vertices are right angles and the angle-sum of the triangle exceeds 180°. We shall now prove that the angle-sum of every triangle exceeds 180°, first establishing the following special case.

Theorem 37.   *The angle-sum of every right triangle exceeds* 180°.

Proof.   The theorem clearly holds for a right triangle with no acute angle or just one. Hence we shall consider a right triangle ABC (Fig. VII, 20) with two acute angles, at A and B, and show that their sum exceeds 90°.

---

* As usual, in a quadrilateral $ABCD$ we call A, C *opposite vertices*, and likewise B, D.

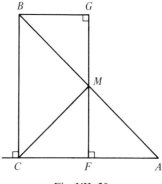

Fig. VII, 20

Let M be the midpoint of $\overline{AB}$. Since line MC subdivides $\angle ACB$ (Ax. 10), $\angle MCA$ is acute. The line through M perpendicular to line AC meets side AC of triangle ACM in an interior point F (Theo. 33). We thus obtain right triangle AFM, in which the length of $\overline{FM}$ is less than $\pi k/2$ (Theo. 31). Extend $\overline{FM}$ in a straight line beyond M to G so that $MG = FM$. The length of $\overline{MG}$ being less than $\pi k/2$, the rectilinear arc FMG consisting of $\overline{FM}$ and $\overline{MG}$ is of length less than πk and hence is a segment shorter than a half-line. Therefore G is on the same side of line AC as B and M, and it is not opposite M [Theo. 30, part (a)]. Since G could be on line BM only if it were opposite M, we see that B, G, M are noncollinear. The triangle of which they are therefore the vertices is congruent to triangle MAF by side–angle–side. Hence $\angle G = 90°$ and $\angle MBG = \angle MAF$.

Next, let P be the pole of line AC on the same side of this line as B. Since $BC < \pi k/2$ (Theo. 31), P is not on $\overline{BC}$. Presently we shall show that $FG < \pi k/2$. Hence P is also not on $\overline{FG}$. As a result, B, C, F, G are the vertices of a Lambert quadrilateral with right angles at C, F, G. The fourth angle, $\angle CBG$, is therefore obtuse (Theo. 35). Points C, F are on the same side of line AB [Theo. 30, part (d)] and F, G are on different sides (Ex. 6 preceding Theo. 31). Hence C, G are on different sides of this line, with the result that the line subdivides $\angle CBG$ (Ex. 7 preceding Theo. 31). Thus we have

$$90° < \angle CBG = \angle CBA + \angle ABG = \angle CBA + \angle CAB,$$

which shows that the acute angles of triangle ABC add up to more than $90°$. The angle-sum of this triangle therefore exceeds $180°$.

It remains to show that $FG < \pi k/2$. Assume $FG = \pi k/2$. Then $G = P$ (Fig. VII, 21).* Since $\angle MBG$, being equal to $\angle MAF$, is acute, its supplement,

---

* It is now convenient to show sides AB, BC, AC as arcs of great circles on a sphere.

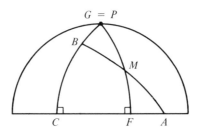

Fig. VII, 21

$\not\prec ABC$, must be obtuse. But this contradicts the original hypothesis that $\not\prec ABC$ is acute. Now assume $FG > \pi k/2$ (Fig. VII, 22).* Then, since $FM = MG$ by an early step in the proof, $FM > MP$. From this fact concerning triangles AMF, BMP, and the further facts $BM = AM$ and $\not\prec AMF = \not\prec BMP$, we conclude that $\not\prec FAM > \not\prec MBP$ (see Ex. 4 following Theo. 22). Hence $\not\prec MBP$ is acute and $\not\prec ABC$, its supplement, is obtuse. This contradicts that $\not\prec ABC$ is acute. We conclude that $FG < \pi k/2$.

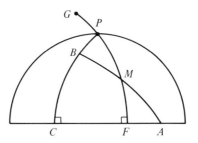

Fig. VII, 22

Theorem 38. *The angle-sum of every triangle exceeds 180°.*

Proof. The theorem clearly holds for a triangle with no acute angle or just one. Hence we need only consider triangles ABC in which at least two angles, say at A and B, are acute. The line through C perpendicular to line AB meets side AB in an interior point D (Theo. 33). Since line CD subdivides $\not\prec ACB$ (Ax. 10), we obtain two right triangles ACD, BCD such that $\not\prec ACD + \not\prec BCD = \not\prec ACB$. By Theorem 37 the sum of the six angles of these triangles exceeds 360°. Since the sum of four of these angles equals the angle-sum of triangle ABC and the sum of the other two, $\not\prec ADC$ and $\not\prec BDC$, is 180°, it follows that the angle-sum of triangle ABC exceeds 180°.

* *G* has been placed where it can be most easily seen.

EXERCISES

1. Show that the angle-sum of a convex quadrilateral exceeds 360°. (See Ex. 6 preceding Theo. 37.)

2. Prove that two triangles are congruent if the angles of one are equal, respectively, to the angles of the other.

VIII

Single Elliptic Geometry

1. INTRODUCTION

Following the same general plan that we used for double elliptic geometry, we shall first try in an informal way to impart an understanding of single elliptic geometry and then proceed systematically to develop the subject from axioms. Single elliptic geometry, we recall, resembles double elliptic geometry in that straight lines are finite and there are no parallel lines, but differs from it in that two straight lines meet in just one point and two points always determine only one straight line. In this latter respect, of course, single elliptic geometry resembles Euclidean and hyperbolic geometry. We mentioned in the preceding chapter that single elliptic geometry is sometimes called elliptic geometry of the hemispherical type because of its relation to the geometry on a hemisphere. Actually, for this statement to be entirely correct the hemisphere needs to be modified in a certain way, which we shall now describe.

2. GEOMETRY ON A MODIFIED HEMISPHERE

Let us take a Euclidean hemisphere H, including its boundary c, and modify it by regarding each two antipodal points of the boundary as the same point (Fig. VIII, 1). This modification is an abstract procedure which cannot be achieved physically, but it is precise and has definite implications, as will be seen presently. We shall refer to the surface resulting from the modification as a *modified hemisphere* and denote it by M. Familiarity with this surface, which is not difficult to acquire because of its close relation to a hemisphere, offers one a good preparation for grasping the main facts of single elliptic

236

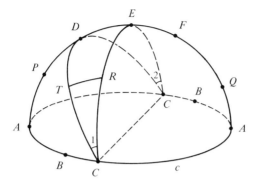

Fig. VIII, 1

geometry. In the discussion to follow we shall denote the sphere containing H by S, its radius by r, and the pole of c situated in H by E (Fig. VIII, 1).

Perhaps the most striking effect of the specified *identification** of antipodal points on c is in the way it changes distance relations. Before the identification the distance between two antipodal points equals half the length of c; after the identification their distance is zero. Points such as P and Q (Fig. VIII, 1) are close after the identification since each is close to A, but they are comparatively far apart before the identification.

The identification also has important effects on the structure of figures and two such effects should be noted immediately. First, the semicircumferences of S situated in H are changed from open curves into closed ones, their lengths retaining the value πr. We shall call these closed curves, such as *AEA*, *CDC*, *CEC* in Fig. VIII, 1, *modified curves*. Second, the boundary c, which is a circle of length $2\pi r$ on H, is changed into a curve, still closed, but whose length must be reckoned as πr since the curve is completely traversed by going continuously from any point of c to the antipodal point. We shall also regard this curve as a modified curve. Thus *all modified curves are simple closed curves and they have the same length* πr. The importance of these curves for our purpose lies in the fact that *the shortest path (on M) joining two points of M is an arc of a modified curve.* This follows, of course, from the close relation between these curves and the great circles on S.

Other important properties of modified curves are to be noted. *Through each point of M there pass infinitely many modified curves, the totality of whose points constitutes M.* This property, which is obvious if the point is E, is a consequence of familiar facts concerning the great circles of S which pass through the point. *Through each two points of M there passes a unique modified*

---

* That is, regarding two antipodal points on c as identical, or the same.

curve. This is because each two points of *M* are nonantipodal points on *S* and a unique great circle passes through two such points on *S. Any two modified curves meet in a unique point.* This is apparent in case one of the curves is *ABA.* Thus, this curve meets the modified curve *CEC* only in *C.* More generally, the specified property is a consequence of the fact that when two great circles of *S* do not meet on *c*, just one of their intersections lies on *H. A modified curve does not divide the rest of M into two separate regions* despite the fact that it is a simple closed curve. For example, it might seem that the modified curve *CEC* divides the rest of *M* into two parts, one containing point *D* and the other point *F*, and that it is not possible to proceed from *D* to *F* along a continuous path without encountering a point of the curve *CEC.* This is not the case, however, for we can proceed downward from *D* to *A*, and then upward from *A* to *F.* The situation is different on a sphere or on an unmodified hemisphere, where every simple closed curve (except the boundary curve of the unmodified hemisphere) will divide the rest of the surface into two separate parts. These surfaces are therefore said to be *two-sided*, whereas a modified hemisphere is called a *one-sided* surface.

The important metric facts concerning *M* and its modified curves remain to be mentioned. *The area of M is* $2\pi r^2$. As on a sphere, a shortest path between two points of *M* is called a *geodesic arc* and the length of this arc is called the *distance* between the points. If the points are *A*, *B* we may denote the arc by $\overarc{AB}$ and its length by *AB.* [It should be noted that if the points are close enough to the boundary curve *c*, their shortest path may cross *c*. Thus, for *P* and *Q* this path is the arc *PAQ* (Fig. VIII, 1).] Two points are called *opposite* if they divide the modified curve through them into equal parts. For example, *C* and *E* are opposite. Hence *opposite points can be joined by exactly two geodesic arcs, nonopposite points by exactly one.* Since the length of a modified curve is πr, *the distance between two opposite points is* $\pi r/2$, *which is the maximum distance possible on M.*

Further, we note that each point of the modified curve *ABA*, but no other point, is opposite *E*, and that the modified curves perpendicular to *ABA* meet in *E.* This illustrates the following fact: *All modified curves perpendicular to a given modified curve meet in a unique point and the given curve is the locus of points at the distance* $\pi r/2$ *from that point.* The point is called the *pole* of the given modified curve and the distance $\pi r/2$ is called *polar distance.* It should be noted that polar distance is half the length of a modified curve, whereas in spherical geometry polar distance is one-fourth the length of a great circle. Also, in spherical geometry, polar distance and maximum distance are not equal, as they are on *M.*

When two modified curves meet in a point not on *c*, the four angles thus formed are simply the angles formed at the point by the corresponding great circles of *S.* Hence the basic facts concerning the angles at such a point are the same as in spherical geometry; that is, the vertical angles are equal, the sum of the four angles is 360°, through such a point on one modified curve there passes just one other modified curve which is perpendicular to it, and so forth. These facts also hold when the curves meet in a point of c. Thus $\measuredangle 1$ and $\measuredangle 2$ in Fig. VIII, 1 are vertical angles and they are equal.

By a *modified hemispherical triangle* or, more briefly, a *hemispherical triangle*, we shall mean a figure on M consisting of three points not on the same modified curve and three geodesic arcs joining them in pairs. The points and arcs are called the *vertices* and *sides* of the triangle. The *angles of a hemispherical triangle* are the three angles, each less than 180°, formed at the vertices by the pairs of sides. Figure VIII, 1 shows several hemispherical triangles, including one with vertices A, C, D and sides ABC, CTD, DPA, and more than one with vertices A,C,E. The three-sided figure $CERTDC$ is not a hemispherical triangle since two of its sides are not geodesic arcs on M. It should be noted, however, that these two sides are geodesic arcs on S, and hence that the figure is a triangle on S, that is, a spherical triangle. The hemispherical triangle ACD mentioned above is also a spherical triangle.

Hemispherical triangles which are also spherical triangles have exactly the same properties as spherical triangles (angle-sum, area, congruence relations, trigonometric formulas, etc.). Hemispherical triangles which do not meet c, or which meet c only in a vertex, are always of this sort. Those which meet c in some other way may fail to be spherical triangles. Consider, for example, points A, B, E in Fig. VIII, 1. They are the vertices of several hemispherical triangles, two of which are shown by heavy lines in Fig. VIII, 2. One of these, t_1, has sides AB, BE, ADE, and the other, t_2, has sides AB, BE, AFE. Although the sides of t_1 are equal, respectively, to those of t_2, t_1 is a spherical triangle but t_2 is not. Angle E in t_2 is the supplement of $\measuredangle E$ in t_1. Hence the triangles are not congruent despite the equality of their sides (also despite the fact that the conditions of side–angle–side and angle–side–angle are met).

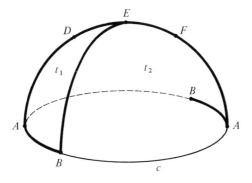

Fig. VIII, 2

Since t_1 is a spherical triangle, its $\measuredangle E$ can be calculated by substituting the lengths of its sides in the spherical Law of Cosines. It follows that $\measuredangle E$ of t_2 cannot be so calculated and hence that t_2 does not obey all the trigonometric formulas of spherical geometry.

Noting that A, E and B, E in the preceding example are pairs of opposite points in M (and that $\overset{\frown}{AE}$, $\overset{\frown}{BE}$ are therefore as long as sides in hemispherical triangles can be), one might ask whether hemispherical triangles would be congruent under the familiar conditions of side–side–side, side–angle–side, and so on, if no two vertices in either triangles were opposite. The answer is in the negative, as we now show.

For simplicity let us take $r = 1$, in which case the length of a modified curve is π. Consider points A, B, D (Fig. VIII, 3) on c such that $AB = AD = BD = \pi/3$, and let F be the midpoint of $\overset{\frown}{AB}$. If C is taken very close to D on

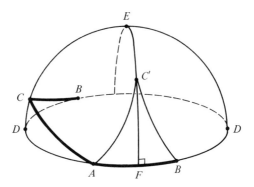

Fig. VIII, 3

the modified curve DE, we obtain a hemispherical triangle ABC (shown by heavy lines) in which sides AC, BC have equal lengths slightly greater than $\pi/3$, $\measuredangle C$ is slightly less than $180°$, and no two vertices are opposite. Figure VIII, 4 gives an additional view of this triangle, which is clearly not a spherical triangle. Let us now choose C' on $\overset{\frown}{EF}$ (Fig. VIII, 3) so that $AC' = AC$. This is possible since the length of $\overset{\frown}{AC}$ is very close to $\pi/3$ and the distance from A to the points of $\overset{\frown}{EF}$ takes on all values from $\pi/6$, which is AF, to $\pi/2$, which is AE. Since $\overset{\frown}{EF}$ is the perpendicular bisector of $\overset{\frown}{AB}$, we know that $AC' = BC'$, and hence that $BC' = BC$. Thus we obtain a hemispherical triangle ABC' whose sides are equal, respectively, to those of hemispherical triangle ABC, and hence whose pairs of vertices are nonopposite. Triangle ABC', being also a spherical triangle, obeys the trigonometric formulas of spherical geometry.

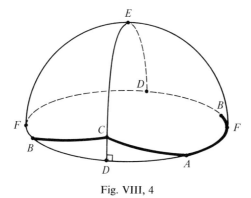

Fig. VIII, 4

In particular, since this triangle is almost equilateral, the computation of $\angle C'$ by the spherical Law of Cosines would give a value that differs only slightly from those of $\angle BAC'$, $\angle ABC'$, and hence is acute. But, as we saw above, $\angle C$ is close to $180°$. The two triangles are therefore not congruent.

Thus we see that two hemispherical triangles, even when no two vertices in either triangle are opposite, may not be congruent when their three sides are equal, respectively. When hemispherical triangles are sufficiently small, however, it can be shown that they are congruent under precisely the same conditions as spherical triangles and obey all the formulas of spherical trigonometry.

EXERCISES

1. Prove that through each point of M there pass infinitely many modified curves, the totality of whose points constitutes M.

2. Prove that all modified curves perpendicular to a given modified curve meet in a unique point and that the given curve is the locus of points at the maximum distance from this point.

3. Prove that any given point of M is the pole of some modified curve.

4. A modified curve is a simple closed curve which does not divide M into two separate parts. Are there any simple closed curves on M which divide it into two separate parts? Justify your answer.

5. Define a circle on M. Does a modified curve satisfy this definition?

6. Draw a circle on M with center A (Fig. VIII, 1) and radius approximately half of the maximum distance. What happens to this circle when its radius approaches the maximum distance?

7. Do two modified curves ever bisect each other? Justify your answer.

8.   State a condition for the congruence of two hemispherical triangles.

9.   Given that the following hemispherical triangles are also spherical triangles, compute the specified parts (use $r = 1$): (a) the hypotenuse of a right triangle each arm of which is one-fourth the length of a modified curve; (b) the angles of a triangle each side of which is one-fifth of the maximum distance on M.

10.   Assuming that sides AC, BC in Fig. VIII, 4 have the length 1.1, show that $\measuredangle ACB = 153°$.

11.   State how many hemispherical triangles there are in Fig. VIII, 1 with the following vertices: (a) C, E, F; (b) A, P, Q; (c) C, R, T; (d) E, R, T.

12.   Verify that each angle of hemispherical triangle ABC in Figs. VIII, 3, and VIII, 4 can be subdivided by modified curves which do not meet the opposite side.

13.   Figure VIII, 5 shows a hemispherical triangle PQR with vertex R on c and side PQ cutting c at A. Verify that the triangle is not also a spherical triangle and that modified curves which subdivide $\measuredangle R$ (shown by arrow) do not meet the opposite side.

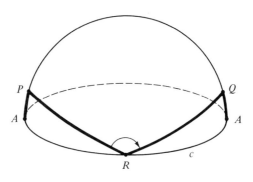

Fig. VIII, 5

14.   In Euclidean and hyperbolic geometry, if three points A, B, C are encountered in this order when a straight line is traversed, then $AB + BC = AC$. Show that this is not always true on a modified curve.

3.   A DESCRIPTION OF SINGLE ELLIPTIC GEOMETRY

The concept of a two-dimensional system of geometry which, like Euclidean geometry, uses the terms point, straight line, plane, circle, angle-size, perpendicular, congruent, distance, and so forth, and whose points and straight lines have the same properties and relations as the points and modified curves of M, is the concept of single elliptic geometry. The facts of this system can therefore be written down immediately by making simple changes in the statements concerning the modified hemisphere. These changes, to mention only the most important, consist of replacing the term "modified curve" by the term "straight line," "geodesic arc" by "straight line segment," "hemispherical triangle" by "triangle," "modified hemisphere" by "single elliptic plane," and r by k in each formula.

We list below some of the basic facts of single elliptic geometry obtained by this procedure. No attempt has been made to classify them as axioms, definitions, and theorems. Each fact bears the same number as the corresponding fact in our list for double elliptic geometry (VII, §5), which will facilitate comparison.

Corresponding Terms

Geometry on a Modified Hemisphere	Single Elliptic Geometry
modified curve	straight line
geodesic arc	straight line segment
hemispherical triangle	triangle
modified hemisphere	single elliptic plane
r	k

Some Basic Facts of Single Elliptic Geometry

1.   Each pair of straight lines meet in exactly one point.

2.   Through each pair of points there passes exactly one straight line.

3.   Through each point there pass infinitely many straight lines, the totality of whose points constitutes the single elliptic plane.

4.   Associated with each two points A, B is a positive number, denoted by AB or BA, called the *distance* between the points. If A, B, C are any three points, then $AB + BC \geq AC$.

5.   The distance between any two points A, B equals the length of a part of the straight line through them. This part of a straight line is called a *straight line segment AB* or, briefly, a *segment AB*. There may be more than one segment AB, but not more than two. A segment joining two points is shorter than any path joining them which is not a segment.

6.   Once a unit segment has been chosen, distances in the single elliptic plane may attain, but not exceed, a certain number of such units. This number is called the *maximum distance* in the plane. Two points at the maximum distance are called *opposite points*, each point being regarded as opposite the other. There are infinitely many points opposite any given point and they form a straight line. Two points always have one common opposite point.

7. A unique straight line goes through any two points. A unique segment connects them if they are nonopposite, and exactly two segments connect them if they are opposite.

8. A pair of opposite points divide their straight line into two equal parts, which are segments. Conversely, if a pair of points divide their straight line into two equal parts, the points are opposite. All straight lines have a length equal to twice the maximum distance.

9. If the maximum distance is denoted by Δ, then the length of a straight line is 2Δ and the area of the single elliptic plane is $8\Delta^2/\pi$. It is customary to express these formulas differently, in terms of the letter k, representing a certain part (about two-thirds) of the maximum distance. Specifically, if we let $k = 2\Delta/\pi$, then the maximum distance is $\pi k/2$, the length of a straight line is πk, and the area of the single elliptic plane is $2\pi k^2$. If we imagine a straight line divided into π equal parts, then k is the length of each part. The value of k depends on the choice of the unit segment inasmuch as the value of Δ does.

10. All straight lines perpendicular to any given straight line meet in the same point. This point, called the *pole* of the line, is at the maximum distance $\pi k/2$ from each point of the line. Also, every point in the plane at the maximum distance from the pole is on the line. The maximum distance is therefore also called *polar distance*.

11. A *triangle ABC* is the figure consisting of three noncollinear points A, B, C and three segments AB, AC, BC. If two of the points are opposite, then (since there are two segments joining them) there is more than one triangle ABC. The definition of congruent triangles is the same as in Euclidean geometry.

12. When triangles are sufficiently restricted in size, their properties are the same as in double elliptic geometry; that is, the facts about angle-sum and area, the conditions for congruence, the trigonometric formulas, and so forth, are the same. If this restriction is not observed, the facts can be quite different. Two triangles, for example, may fail to be congruent even when the conditions of side–side–side, side–angle–side, angle–side–angle, angle–angle–side are met, and a straight line which subdivides an angle of a triangle may fail to meet the opposite side. On the other hand, there exist three-sided rectilinear figures whose sides are longer than segments, but whose other basic properties (angle-sum, area, conditions for congruence, trigonometric formulas, etc.) are the same as those of triangles restricted in size. (An example of a figure on a modified hemisphere corresponding to such a three-sided rectilinear figure in the single elliptic plane is the three-sided figure *CERTDC* in Fig. VIII, 1.)

The relation between single elliptic plane geometry and the geometry on a modified hemisphere can now be described more precisely than was done at the beginning of this section. Let such a hemisphere, with an arbitrary but specific value of r, be given and let a unit segment in the single elliptic plane be chosen so that $k = r$. Then the geometry on the modified hemisphere and single elliptic plane geometry are logically indistinguishable. Except for differences in certain words and symbols, the statements and formulas in one system are identical with the statements and formulas in the other.

EXERCISES

All exercises refer to single elliptic geometry except when otherwise indicated.

1.   Define a circle and verify that straight lines are circles.

2.   The circumference of a circle is less than 2π times its radius. Verify this for a circle with radius $\pi k/2$.

3.   Describe the locus of the points whose distance from a given point does not exceed $\pi k/2$.

4.   Prove that a circle with a radius less than $\pi k/2$ is not a straight line.

5.   The interior of a circle not a straight line is defined as the set of all points whose distance from the center of the circle is less than the radius. (a) Define the exterior of such a circle. (b) Why does it not make sense to define the interior of a circle which is a straight line?

6. The formulas for the circumference C and area S of a circle of radius a are the same as in double elliptic geometry when $a < \pi k/2$ (see VII, §5, Exs. 5, 7). Show that only one of these formulas holds when $a = \pi k/2$ and account for the difference.

7.   What can be said about the behavior of the formulas for the circumference and area of a circle when the radius approaches zero? (See Ex. 6 and VII, §5, Exs. 12, 13.)

8.   (a) If a triangle is restricted in size so that its properties are the same as those of a triangle in double elliptic geometry, what can be said about its angle-sum when its area approaches zero? (See VII, §5, Ex. 11.) (b) Under these same circumstances, what can be said about its trigonometric relations?

9.   If A, B, C are distinct noncollinear points in double elliptic geometry, then no two of them are opposite. Is this true in single elliptic geometry? Justify your answer.

10.   In Euclidean geometry three collinear points can always be labeled A, B, C so that $AB + BC = AC$. Is the same true in single elliptic geometry? Justify your answer.

11.   If a modified hemisphere with the value $r = 6/\pi$ is given, how should the unit segment be chosen in the single elliptic plane so that $k = r$?

4.   AN AXIOMATIC PRESENTATION OF SINGLE ELLIPTIC GEOMETRY

Our objective in the preceding two sections was to acquaint the student with some of the basic facts of single elliptic geometry and to show how these facts can be determined from a knowledge of the geometry on a modified hemisphere. Although the facts were stated in a certain logical order, no attempt was made to shape them into a deductive system based on axioms, and no proofs were given. In the present section we wish to make some effort in this direction. As in our axiomatic presentation of double elliptic geometry, the procedure again will be simply to prove some facts on the basis of others. The latter, which will not be proved, will therefore serve as our axioms and definitions, and the proved facts will be our theorems. We shall not go as far as we did with double elliptic geometry.

Familiarity with all the terms (metric space, simple arc, simple closed curve, etc.) and ideas used in the axiomatic presentation of double elliptic geometry is presumed. Again we shall not go into details concerning length, but take its familiar properties (VII, §7) for granted. The distance between two points A, B will once more be denoted by AB or BA and the term "simple arc AB" will mean a simple arc with endpoints A, B. As in our development of double elliptic geometry, "point" and "straight line" will be undefined terms. Not all of the axioms are stated at once. The presentation is divided into several parts.

Straight Lines and Segments

Axiom 1.   *The single elliptic plane is a metric space containing at least two points, at least one simple arc of finite length joining each two points, and at least two straight lines.*

Axiom 2.   *A straight line is a simple closed curve of finite length.*

These axioms, of course, are not definitions of a straight line and of a single elliptic plane although their wording might suggest this. They merely state a few basic properties which we are assuming about the line and the plane, and which will be useful in deducing others.

Since simple arcs and simple closed curves are infinite sets of points, Axioms 1 and 2 imply

Theorem 1. *The single elliptic plane contains infinitely many points and not all of them are on the same straight line.*

Axiom 3. *Each two straight lines meet in exactly one point.*

From Axiom 3 we obtain immediately

Theorem 2. *Not more than one straight line can go through two points.*

This theorem merely states that there cannot be as many as two straight lines through a pair of points. We do not know yet that there *is* a straight line through the points. Later we shall prove that there is always exactly one straight line through them.

Axiom 4. *Among the simple arcs joining each two points there is one (or more) whose length is least, that is, no other simple arc joining the points has a smaller length. The length of this simple arc (or each of them if there is more than one) equals the distance between its endpoints.*

Definition 1. A simple arc with the property stated in Axiom 4 is called a *segment*. If there is exactly one segment joining two points, we shall refer to it as *the segment* joining the points and call it *unique*. The term *segment AB* will denote a segment with endpoints *A*, *B*.

Axiom 5. *At least one of the simple arcs into which a straight line is divided by each two of its points is a segment.*

From Definition 1 and Axioms 4 and 5 we deduce

Theorem 3. *If the two simple arcs into which a straight line is divided by a pair of its points are unequal (in length), only the shorter is a segment joining the points; if they are equal, both are segments joining the points. The length of a segment AB is equal to the distance AB.*

Definition 2. Two points which divide a straight line into equal arcs are called *opposite points on the line* (or simply *opposite points*), and each point is said to be *opposite the other on the line* (or simply *opposite the other*). The two arcs are called *half-lines*. The term *half-line ACB* will mean a half-line with endpoints *A*, *B* and interior point *C*.

It is useful to combine Theorem 2 and Definition 2 as follows.

Theorem 4.   *The equal arcs into which each two opposite points on a straight line divide the line are segments joining the points.*

Axiom 6.   *Each segment joining two points is (a simple arc) on some straight line through the points.*

If A, B are any two points, there is at least one segment joining them according to Axiom 4 and Definition 1, and there is a straight line through them which contains this segment by Axiom 6. There cannot be a second straight line through them by Theorem 2. Hence we have

Theorem 5.   *Exactly one straight line goes through each two points and it contains all the segments joining the points.*

In view of Theorem 5, if B is opposite A on one straight line through A, and C is opposite A on another straight line through A, then B and C are distinct, unlike the situation in double elliptic geometry. In other words, there is no unique point which is opposite a given point. Later we shall prove that the locus of points opposite any point is a straight line.

Another important difference can be noted at this time. In double elliptic geometry, if three points are noncollinear, then no two of them are opposite. This is not always true in single elliptic geometry. For let A, B be a pair of opposite points. Since there is exactly one straight line through them (Theo. 5) and not all points of the single elliptic plane are on the same straight line (Theo. 1), there must be a point C which is not on straight line AB. Thus A, B, C are noncollinear though A, B are opposite.

Now let A, B be any two points. If they are opposite on the straight line through them, then they divide the line into two equal segments (Theo. 4) and these are the only segments joining the points (Theo. 5). If A, B are non-opposite on their straight line, then only the shorter of the two arcs into which they divide the line is a segment (Theo. 3) and it is the only segment joining them (Theo. 5). Thus we have

Theorem 6.   *The number of segments joining two points is one or two according as the points are nonopposite or opposite on the straight line through them.*

Rays, Angles, and Triangles

Definition 3.   The half-line ACB exclusive of B is called the *ray AC* and exclusive of A is called the *ray BC*. The *endpoint* of ray AC is A and that of ray BC is B; in each case C is an *interior point* of the ray. The figure formed by two rays with a common endpoint is called an *angle*; the endpoint and rays are called the *vertex* and *sides* of the angle. If the two sides of an angle are collinear, the angle is called a *straight angle*. Consider the figure formed by two simple arcs PQ, PR such that each arc is on a straight line and the arcs have only P in common (Fig. VIII, 6). Of the two rays on line PQ

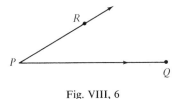

Fig. VIII, 6

with endpoint P choose the one which is traversed by proceeding along the arc PQ starting from P. Similarly, choose the ray on line PR with endpoint P which is traversed by proceeding along the arc PR starting from P. (Figure VIII, 6 illustrates the case in which the ray on line PQ is shorter than arc PQ and the ray on line PR is longer than arc PR. The rays are represented by arrows.) We shall call the angle consisting of these two rays the *angle of the figure formed by the two arcs PQ, PR* and denote it by $\measuredangle RPQ$ or $\measuredangle QPR$.

The basic facts and terminology (acute angle, obtuse angle, right angle, vertical angles, perpendicular, etc.) relating to the measurement and naming of angles are the same as in Euclidean and double elliptic geometry. The angular theory at a point, as this term was defined in Chapter VII, Section 7, is also essentially the same. In our presentation of double elliptic geometry we avoided the lengthy discussion needed to prove this theory by accepting it as an axiom and we shall now do likewise for that small part of the theory used in our work.

Axiom 7. *Vertical angles are equal. All right angles are equal. At each point of a straight line there is a unique straight line which is perpendicular to the line.*

In double elliptic geometry, in order to define an interior point and a subdivider of an angle we used the fact that two points on different sides of an angle other than a straight angle are nonopposite. In single elliptic geometry, however, two such points can be opposite, as is to be shown later in an exercise, and so an interior point and a subdivider must be defined differently. It may be of interest to give the definitions even though we shall have no occasion to use them in this brief presentation of single elliptic geometry. A *subdivider of an angle ABC* other than a straight angle is any ray *BD* not coincident with a side of the angle such that $\angle ABC = \angle ABD + \angle CBD$ and an *interior point* of the angle is any inner point of a subdivider. We leave it to the student to define a subdivider of a straight angle.

Definition 4. The figure consisting of three noncollinear points *A, B, C* and three segments *AB, AC, BC* is called a *triangle ABC*. The points and segments are called the *vertices* and *sides* of the triangle. The *angles of the triangle* are the three angles *ABC, BCA, CAB*. Two triangles are called *congruent* if the sides and angles of one are equal, respectively, to the sides and angles of the other.

Clearly, no two sides of a triangle are collinear. Hence no angle of a triangle is a straight angle. Each angle is therefore less than 180°. Since two (perhaps each two) vertices may be opposite points, a side may not be the only segment joining its endpoints. It follows that there may be more than one triangle with given vertices, unlike the situation in double elliptic geometry. This is one reason why the familiar propositions on the congruence of triangles do not hold without some restriction.

In our development of double elliptic geometry one of these familiar facts about congruent triangles was accepted as an axiom. For the reason given above, this cannot be done in single elliptic geometry. Yet some assumptions concerning congruence must be made if we are to proceed further. Since we cannot make them about triangles, we shall make them for a type of figure, slightly more general than a triangle, which we shall call a *three-sided rectilinear figure*. It differs from a triangle only in that its sides are not necessarily segments, but may be any rectilinear simple arcs, that is, any simple arcs of straight lines. The formal definition of this figure, and of the related terms vertices, sides, angles, congruent, can therefore be obtained from Definition 4 merely by replacing "triangle" by "three-sided rectilinear figure" and "segments" by "rectilinear simple arcs." A triangle, then, is a three-sided rectilinear figure all of whose sides are segments.

Concerning the new figure we make two assumptions:

Axiom 8. *If two angles of a three-sided rectilinear figure are right angles, the sides opposite them are equal.*

Axiom 9. *A pair of three-sided rectilinear figures, each containing two right angles, are congruent if the sides included by these angles are segments and the remaining angles are equal.*

Pole and Polar

The next theorem states an important fact concerning the perpendiculars to a straight line. Before proving the theorem let us agree to speak of the two simple arcs into which a pair of points divide a straight line as being *complements* of each other.

Now let p be any straight line and let the straight lines perpendicular to it at two of its points M, N meet in point P. From among the various three-sided rectilinear figures MNP let us choose one whose side MN is a segment, denote it by f, and exhibit it as in Fig. VIII, 7. We note that f is the type of

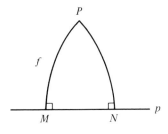

Fig. VIII, 7

figure mentioned in Axioms 8 and 9. Another example of that type is the three-sided rectilinear figure MNP whose side MN is the same as side MN in f and whose sides MP, NP are the complements of sides MP, NP in f. We denote this figure by f' and exhibit it conveniently as in Fig. VIII, 8. Sides MP, NP

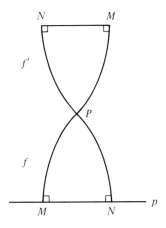

Fig. VIII, 8

in f' can be regarded as the extensions, respectively, of sides MP, NP in f beyond P. Thus $\not\!\prec MPN$ in f and $\not\!\prec MPN$ in f' are vertical angles, and therefore equal (Ax. 7). Hence f and f' are congruent by Axiom 9. This implies that the sides opposite the right angles in f are equal, respectively, to the sides opposite the right angles in f'.

Also, by Axiom 8 sides MP, NP in f are equal, and the same is true of sides MP, NP in f'. Thus the four sides MP, NP are equal. The two sides MP, being equal and complementary, are segments (Theo. 3) and M, P are opposite (Def. 2). Similarly, the two sides NP are segments and N, P are opposite. The four segments MP, NP being equal, P is equidistant from M and N (Theo. 3).

What has been shown thus far is that the straight lines perpendicular to p at M, N meet in a point P which is opposite M and N, and equidistant from them.* Since M, N are any two points of p, these results can be generalized. To do this let us think of M as fixed and N as variable. From what has been shown it then follows that regardless of the position of N, the straight line perpendicular to p at N must meet straight line MP in the point of the latter which is opposite M, that is, in P. Thus *all* the straight lines perpendicular to p meet in P, and P is opposite and equidistant from *all* the points in which those lines meet p. Conversely, let b be any straight line through P. Then b meets p in a point B. The straight line perpendicular to p at B must go through P, as shown above, and hence must be identical with b (Theo. 5). Thus each straight line through P is perpendicular to p. The proof of the following is now complete.

Theorem 7.   *All straight lines perpendicular to any given straight line p go through the same point P and all straight lines through P are perpendicular to p, meeting it in points all of which are opposite P and equidistant from it.*

Definition 6.   The point P in the preceding theorem is called the *pole* of p, p is called the *polar* of P, and the constant distance from P to the points of p is called the *polar distance* for p.

In proving Theorem 7 we started out with any straight line p and then determined the point P. Hence we showed that every straight line has a pole. It is also true that every point is the pole of some straight line. For let A be any point, g any straight line through A, B the point of g opposite A, and a the straight line through B perpendicular to g. Since g is the straight line perpendicular to a at B, the pole of a is the point on g opposite B. Thus A is the pole of a. We can therefore state

---

* Incidentally, we have also shown that a three-sided rectilinear figure is necessarily an isosceles triangle if one of its sides is a segment included by right angles.

Theorem 8.   *Every straight line has a pole and every point is the pole of some straight line* (*that is, every point has a polar*).

From Theorems 7 and 8 and Definition 6 we obtain

Theorem 9.   *The set of straight lines through any point is identical with the set of straight lines perpendicular to the polar of the point.*

If P is any point, then each point on its polar p is opposite P (Theos. 7, 8 and Def. 6). There can be no other point which is opposite P, for suppose Q were such a point. Straight line PQ would then meet p in a point R which is opposite P (Theo. 7). Thus there would be two points on straight line PQ which are opposite P, namely, Q and R, and this is impossible. We can therefore state

Theorem 10.   *The locus of points opposite any point is a straight line, the polar of the point.*

Maximum Distance and Length of a Straight Line

It is clear from the discussion which led to Theorem 7 that if we let Δ denote the polar distance for the straight line p, then 2Δ is the length of each straight line perpendicular to p. To generalize this result, let q and r (Fig. VIII, 9) be two straight lines through P which are perpendicular to each

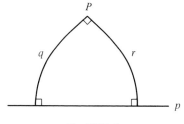

Fig. VIII, 9

other. They are also perpendicular to p (Theo. 7) and hence have the same length 2Δ. Now r, like p, has the property that all straight lines perpendicular to it have a common length. Hence p and q, being perpendicular to r, have the same length. Thus p, as well as each straight line perpendicular to it, has the length 2Δ.

Similarly, if p' is any straight line other than p, P' its pole, and Δ' the polar distance for p', then $2\Delta'$ is the common length of p' and each straight line perpendicular to it. Now, if A, B are two points on p (Fig. VIII, 10) and we choose a segment AB, a segment AP, and a segment BP, we obtain a certain triangle ABP having these segments as sides and with right angles at A and B.

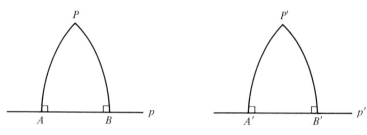

Fig. VIII, 10

A triangle $A'B'P'$ congruent to this one will result, according to Axiom 9, if we take as two of its sides a pair of segments $A'P'$, $B'P'$ such that $\angle APB = \angle A'P'B'$, where A', B' are on p', and as the third side a segment $A'B'$. Sides AP, $A'P'$ are then equal, and so $\Delta = \Delta'$. Hence p and p' have equal lengths. This proves

Theorem 11.   *All straight lines have the same polar distance and a length equal to twice this distance.*

Definition 7.   The numerical value of the common polar distance being dependent on the unit of length, that is, on the unit segment, we follow custom in denoting the value by $\pi k/2$, where k is a parameter which depends on the unit of length.

For example, if a segment joining two opposite points is the unit of length, then the polar distance is 1, so that $\pi k/2 = 1$ and $k = 2/\pi$. If half of this segment is the unit of length, then the polar distance is 2, $\pi k/2 = 2$, and $k = 4/\pi$.

In what follows we assume that some unspecified unit of length has been chosen and hence that k is constant, though unknown. From Theorem 11 and Definition 7 we obtain

Theorem 12.   *All straight lines have the same length πk.*

If P, Q are any two points, they are either opposite or nonopposite on the straight line determined by them. If they are opposite, they divide the line into two equal parts, each being a segment of length $\pi k/2$. Hence the distance between them is $\pi k/2$ (Theo. 3). If P, Q are nonopposite, they divide the line into two unequal parts. The shorter part must have a length less than $\pi k/2$ since the length of the line is πk. Also, the shorter part is a segment (Theo. 3). Hence the distance between P and Q is less than $\pi k/2$ (Theo. 3). We have thus proved

Theorem 13. *There is a maximum distance in the single elliptic plane and its value is $\pi k/2$. This distance is attained by each two opposite points, and only by them.*

We leave the proof of the following as an exercise.

Theorem 14. *Through a point not a pole of a given straight line there passes a unique straight line perpendicular to the given line.*

Definition 8. A *circle* is a locus consisting of all points whose distances from some point are equal. The latter point and the common distance are called the *center* and *radius* of the circle.

The maximum distance in the plane being $\pi k/2$, no circle can have a radius exceeding this value. According to Theorem 10, the locus of points opposite any point is the polar of the point, and, by Theorem 13, the distance $\pi k/2$ is attained only by opposite points. It follows that every point in the plane is the center of a circle of radius $\pi k/2$ and that this circle is a straight line, the polar of the point. We leave it to the student to show that no circle with a radius less than $\pi k/2$ can be a straight line. Thus we can state

Theorem 15. *No circle has a radius exceeding $\pi k/2$. Every point in the plane is the center of a circle of radius $\pi k/2$. A circle of radius $\pi k/2$ is a straight line, namely, the polar of the center of the circle. No circle of radius less than $\pi k/2$ is a straight line.*

EXERCISES

1. Prove Theorem 14.

2. Show that no circle with a radius less than $\pi k/2$ is a straight line.

3. Prove that two points on different sides of an angle other than a straight angle can be opposite.

4. Find a pair of circles which meet in (a) two points, (b) one point, (c) no point.

5.   The *tangent* at a point A of a circle with center C is defined to be the straight line through A perpendicular to the straight line AC. Find two circles which meet in just one point without having a common tangent at the point.

6.   (a) Define the interior and the exterior of a circle. (b) According to your definition does every circle have an interior and an exterior? Justify your answer.

We recall from the preceding chapter (see discussion following Theo. 27) that each straight line g in the double elliptic plane divides the rest of the plane into two parts, called the sides of the line, and that the plane is said to be two-sided. Any rectilinear path (that is, any simple arc of a straight line) joining a point on one side of g to a point on the other side will meet g. As our work with a modified hemisphere suggested, the single elliptic plane does not have these properties and is *one-sided*. The following theorem makes this more precise.

Theorem 16. *If h is any straight line in the single elliptic plane, and A, B are any points not on h, there is a rectilinear path joining A and B which does not intersect h.*

Proof.   Through A and B there passes a unique straight line j, which is divided by these points into two simple arcs s_1 and s_2. Line j meets h in a unique point C which is either on s_1 or s_2, but not on both. If C is on s_1, then s_2 is a rectilinear path joining A and B which does not meet h; and if C is on s_2, then s_1 is such a rectilinear path.

Since A and B can be joined by a rectilinear path not intersecting h, we may say that they lie on the *same side* of h. But A and B are *any* points of the plane not on h. Hence we may say that all points of the plane not on h lie on the same side of h, or that the plane has only one side with respect to h, or that h does not separate the rest of the plane into two parts.

Appendix

A Summary
from Euclid's *Elements**

* Taken, with permission, from T. L. Heath, *The Thirteen Books of Euclid's Elements* (Cambridge University Press, New York, 1926).

THE TEN STATED ASSUMPTIONS

The Postulates

Let the following be postulated:
1. To draw a straight line from any point to any point.
2. To produce a finite straight line continuously in a straight line.
3. To describe a circle with any centre and distance.
4. That all right angles are equal to one another.
5. That, if a straight line falling on two straight lines make the interior angles on the same side less than two right angles, the two straight lines, if produced indefinitely, meet on that side on which are the angles less than the two right angles.

The Common Notions (or Axioms)

1. Things which are equal to the same thing are also equal to one another.
2. If equals be added to equals, the wholes are equal.
3. If equals be subtracted from equals, the remainders are equal.
4. Things which coincide with one another are equal to one another.
5. The whole is greater than the part.

THE DEFINITIONS OF BOOK I

1.   A *point* is that which has no part.*
2.   A *line* is breadthless length.
3.   The extremities of a line are points.
4.   A *straight line* is a line which lies evenly with the points on itself.
5.   A *surface* is that which has length and breadth only.
6.   The extremities of a surface are lines.
7.   A *plane surface* is a surface which lies evenly with the straight lines on itself.
8.   A *plane angle* is the inclination to one another of two lines in a plane which meet one another and do not lie in a straight line.
9.   And when the lines containing the angle are straight, the angle is called *rectilineal.*
10.   When a straight line set up on a straight line makes the adjacent angles equal to one another, each of the equal angles is *right,* and the straight line standing on the other is called a *perpendicular* to that on which it stands.
11.   An *obtuse angle* is an angle greater than a right angle.
12.   An *acute angle* is an angle less than a right angle.
13.   A *boundary* is that which is an extremity of anything.
14.   A *figure* is that which is contained by any boundary or boundaries.
15.   A *circle* is a plane figure contained by one line such that all the straight lines falling upon it from one point among those lying within the figure are equal to one another.
16.   And the point is called the *centre* of the circle.
17.   A *diameter* of the circle is any straight line drawn through the centre and terminated in both directions by the circumference of the circle, and such a straight line also bisects the circle.
18.   A *semicircle* is the figure contained by the diameter and the circum-ference cut off by it. And the centre of the semicircle is the same as that of the circle.
19.   *Rectilineal figures* are those which are contained by straight lines, *trilateral* figures being those contained by three, *quadrilateral* those contained by four, and *multilateral* those contained by more than four straight lines.
20.   Of trilateral figures, an *equilateral triangle* is that which has its three sides equal, an *isosceles triangle* that which has two of its sides alone equal, and a *scalene triangle* that which has its three sides unequal.
21.   Further, of trilateral figures, a *right-angled triangle* is that which has a right angle, an *obtuse-angled triangle* that which has an obtuse angle, and and *acute-angled triangle* that which has its three angles acute.

---

* This definition and several others are extremely vague. Euclid never uses them.

22. Of quadrilateral figures, a *square* is that which is both equilateral and right-angled; an *oblong* that which is right-angled but not equilateral; a *rhombus* that which is equilateral but not right-angled; and a *rhomboid* that which has its opposite sides and angles equal to one another but is neither equilateral nor right-angled. And let quadrilaterals other than these be called *trapezia*.

23. *Parallel* straight lines are straight lines which, being in the same plane and being produced indefinitely in both directions, do not meet one another in either direction.

THE PROPOSITIONS OF BOOK I

1. On a given finite straight line to construct an equilateral triangle.

2. To place at a given point (as an extremity) a straight line equal to a given straight line.

3. Given two unequal straight lines, to cut off from the greater a straight line equal to the less.

4. If two triangles have the two sides equal to two sides respectively, and have the angles contained by the equal straight lines equal, they will also have the base equal to the base, the triangle will be equal to the triangle, and the remaining angles will be equal to the remaining angles respectively, namely those which the equal sides subtend.

5. In isosceles triangles the angles at the base are equal to one another, and, if the equal straight lines be produced further, the angles under the base will be equal to one another.

6. If in a triangle two angles be equal to one another, the sides which subtend the equal angles will also be equal to one another.

7. Given two straight lines constructed on a straight line (from its extremities) and meeting in a point, there cannot be constructed on the same straight line (from its extremities), and on the same side of it, two other straight lines meeting in another point and equal to the former two respectively, namely each to that which has the same extremity with it.

8. If two triangles have the two sides equal to two sides respectively, and have also the base equal to the base, they will also have the angles equal which are contained by the equal straight lines.

9. To bisect a given rectilineal angle.

10. To bisect a given finite straight line.

11. To draw a straight line at right angles to a given straight line from a given point on it.

12. To a given infinite straight line, from a given point which is not on it, to draw a perpendicular straight line.

13. If a straight line set up on a straight line make angles, it will make either two right angles or angles equal to two right angles.

14. If with any straight line, and at a point on it, two straight lines not lying on the same side make the adjacent angles equal to two right angles, the two straight lines will be in a straight line with one another.

15. If two straight lines cut one another, they make the vertical angles equal to one another.

16. In any triangle, if one of the sides be produced, the exterior angle is greater than either of the interior and opposite angles.

17. In any triangle two angles taken together in any manner are less than two right angles.

18. In any triangle the greater side subtends the greater angle.

19. In any triangle the greater angle is subtended by the greater side.

20. In any triangle two sides taken together in any manner are greater than the remaining one.

21. If on one of the sides of a triangle, from its extremities, there be constructed two straight lines meeting within the triangle, the straight lines so constructed will be less than the remaining two sides of the triangle, but will contain a greater angle.

22. Out of three straight lines, which are equal to three given straight lines, to construct a triangle: thus it is necessary that two of the straight lines taken together in any manner should be greater than the remaining one.

23. On a given straight line and at a point on it to construct a rectilineal angle equal to a given rectilineal angle.

24. If two triangles have the two sides equal to two sides respectively, but have the one of the angles contained by the equal straight lines greater than the other, they will also have the base greater than the base.

25. If two triangles have the two sides equal to two sides respectively, but have the base greater than the base, they will also have the one of the angles contained by the equal straight lines greater than the other.

26. If two triangles have the two angles equal to two angles respectively, and one side equal to one side, namely, either the side adjoining the equal angles, or that subtending one of the equal angles, they will also have the remaining sides equal to the remaining sides and the remaining angle to the remaining angle.

27. If a straight line falling on two straight lines make the alternate angles equal to one another, the straight lines will be parallel to one another.

28. If a straight line falling on two straight lines make the exterior angle equal to the interior and opposite angle on the same side, or the interior angles on the same side equal to two right angles, the straight lines will be parallel to one another.

29. A straight line falling on parallel straight lines makes the alternate

angles equal to one another, the exterior angle equal to the interior and opposite angle, and the interior angles on the same side equal to two right angles.

30. Straight lines parallel to the same straight line are also parallel to one another.

31. Through a given point to draw a straight line parallel to a given straight line.

32. In any triangle, if one of the sides be produced, the exterior angle is equal to the two interior and opposite angles, and the three interior angles of the triangle are equal to two right angles.

33. The straight lines joining equal and parallel straight lines (at the extremities which are) in the same directions (respectively) are themselves also equal and parallel.

34. In parallelogrammic areas the opposite sides and angles are equal to one another, and the diameter bisects the areas.

35. Parallelograms which are on the same base and in the same parallels are equal to one another.

36. Parallelograms which are on equal bases and in the same parallels are equal to one another.

37. Triangles which are on the same base and in the same parallels are equal to one another.

38. Triangles which are on equal bases and in the same parallels are equal to one another.

39. Equal triangles which are on the same base and on the same side are also equal in the same parallels.

40. Equal triangles which are on equal bases and on the same side are also in the same parallels.

41. If a parallelogram have the same base with a triangle and be in the same parallels, the parallelogram is double of the triangle.

42. To construct, in a given rectilineal angle, a parallelogram equal to a given triangle.

43. In any parallelogram the complements of the parallelograms about the diameter are equal to one another.

44. To a given straight line to apply, in a given rectilineal angle, a parallelogram equal to a given triangle.

45. To construct, in a given rectilineal angle, a parallelogram equal to a given rectilineal figure.

46. On a given straight line to describe a square.

47. In right-angled triangles the square on the side subtending the right angle is equal to the squares on the sides containing the right angle.

48. If in a triangle the square on one of the sides be equal to the squares on the remaining two sides of the triangle, the angle contained by the remaining two sides of the triangle is right.

Bibliography

Bell, E. T., *The Search for Truth*, Chap. XIV (Reynal and Hitchcock, New York, 1934). Non-Euclidean geometry as a landmark in the history of thought.

Bolyai, J., *The Science of Absolute Space* (included as an appendix in the reference to Bonola listed below).

Bonola, R., *Non-Euclidean Geometry, A Critical and Historical Study of Its Development* (Chicago, 1912; reprinted by Dover, New York, 1955).

Borsuk, K. and W. Szmielew, *Foundations of Geometry* (North-Holland Publ., Amsterdam, 1960). A rigorous development of Euclidean and hyperbolic geometry.

Kulczycki, S., *Non-Euclidean Geometry* (Pergamon, Oxford, 1961).

Lieber, L. R., *Non-Euclidean Geometry, or Three Moons in Mathesis* (Science Press, Lancaster, Pennsylvania, 1940). A very brief, elementary, and entertaining account.

Lobachevski, N. I., *Geometric Researches on the Theory of Parallels* (included as an appendix in the reference to Bonola).

Manning, H. P., *Non-Euclidean Geometry* (Boston, 1901; reprinted by Dover, New York, 1963).

Meschkowski, H., *Non-Euclidean Geometry* (Academic Press, New York, 1964).

Prenowitz, W. and M. Jordan, *Basic Concepts of Geometry* (Ginn (Blaisdell), Boston, Massachusetts, 1965). Discusses, among other things, the logical consistency and empirical validity of the non-Euclidean geometries.

Saccheri, G., *Euclides Vindicatus* (Open Court Publ., Chicago, Illinois, 1920). A translation of Saccheri's famous work.

Smith, D. E., *A Sourcebook in Mathematics* (Dover, New York, 1959). Includes translations of works by Saccheri, Bolyai, Lobachevski, and Riemann.

Young, J. W., *Fundamental Concepts of Algebra and Geometry* (Macmillan, New York, 1930). Discusses an imaginary non-Euclidean world.

Answers
to Selected Exercises

CHAPTER I

Section 4

1. Does not depend on Postulate 5.
3. Does not depend on Postulate 5.

Section 5

9. Can be proved without using Postulate 5.
11. Cannot be proved without using Postulate 5. For the stated fact is not true unless alternate interior angles are always equal, and they are not always equal unless Postulate 5 holds.

CHAPTER II

Section 3

1. Not necessarily since it does not follow that they are everywhere equidistant.

Section 5

3. Yes; no.
6. (a) *Hint*. Assume the contrary cases, use Ex. 5, and reach contradictions.

CHAPTER III

Section 5

1. (a) A line which meets one of two parallels may fail to meet the other.
   (b) Alternate interior angles are not always equal.
   (c) At least one triangle exists whose angle sum is not 180°.
   (d) There are parallel lines having no common perpendicular.
   (e) Two lines may be parallel to the same line without being parallel to each other.
   (f) Three noncollinear points exist such that no circle passes through them.

Section 6

3. The summit is longer than the base.
4. *Hint.* Build such a rhombus by using isosceles right triangles.

Section 7

11. The distance is not constant by Ex. 10. If it increased sometimes and decreased at other times, then (this variation being continuous by §2, Property 15) it would have to have the same value at least twice, contradicting Ex. 10.

Section 9

3. Because the area of a triangle has not been defined yet.
7. Yes, since the defect of a Euclidean triangle is zero.

Section 11

7. The step is the assertion in the fourth sentence that Q and Q' are congruent. In Euclidean geometry two Saccheri quadrilaterals with equal summits and equal summit angles are generally not congruent.

Section 12

1. 40.
7. $A = k(180 - S)$, where the angle-sum is $S°$.

Section 13

3. Their product is 1.
9. The area of a quadrilateral is the sum of the areas of the two triangles into which it is subdivided by a diagonal. More precisely, if the quadrilateral is *PQRS*, its area is the sum of the areas of triangles *PQR* and *PRS*.
11. The defect of a quadrilateral is the positive difference between its angle-sum (in degrees) and 360°.

Section 14

5. Yes.

CHAPTER IV

Section 2

1. Boundary parallel in (a), (d), (f). Nonboundary parallel in (b), (c). Indeterminate in (e).

Section 3

1. (a) Some meet *h*; some do not.
   (b) All meet *h*.   (c) All meet *h*.
   (d) For convenience assume *h* is horizontal and *g* is a left-hand parallel to *h*. Then the boundary rays on *h* are the rays which extend to the left; the boundary rays on *g* are the rays which project into the boundary rays on *h*.
3. (a) False.   (b) False.   (c) True.   (d) False.

Section 5

9. *Hint.* Use the second part of Theo. 15.

Section 6

3. No, since the two common perpendiculars generally have different lengths.

Section 7

1. (a) True.   (b) False.   (c) False.   (d) True.   (e) False.   (f) True.

CHAPTER V

Section 3

3.   No, there are exactly two.

Section 5

6. $\sqrt{5}$.
7. $\sqrt[3]{5}$.

Section 6

3.   About four-tenths of a unit.

Section 7

3.   (a) 0.55;   (b) becomes infinite.
7.   0.48 and 0.11.

CHAPTER VI

Section 2

1.   (a) 30°;   (b) 70°;   (c) $\beta = 40°$;   (d) 0.55;   (e) 6°;   (f) 0.10.
5.   Yes, if $b = c'$, which may occur when $\gamma < 45°$.

Section 5

1.   (a) 0.324, 19°;   (b) 0.133, 8°;   (c) 0.443, 26°.
3.   (a) 0.648, 33°;   (b) 0.266, 15°;   (c) 0.887, 42°.
5.   (a) $a = b = 1.15$, $c = 1.76$;
     (b) $a = b = 0.61$, $c = 0.89$;
     (c) $\lambda = 2°$, $\mu = 6°$, $c = 6.3$.

Section 6

1. $\csc \lambda > c/a$, $\sec \lambda < c/b$, $\cot \lambda > b/a$.
11. (a) 59°40′;   (b) 59°55′;   (c) 44°12′;   (d) 44°58′.

Section 7

1.   (a) 52°38′;   (b) 57°59′;   (c) 59°36′;   (d) 59°58′.

Section 8

2. *Hint.* Use the relation between a Lambert quadrilateral and a Saccheri quadrilateral.
5. The hyperbolic values are:
   (a) 5.524 π; (b) 1.086 π; (c) 0.2552 π; (d) 0.04016 π.

Section 9

1. (a) The angle-sum is between 179°59′59″ and 180°.
   (b) It is less than 180° but greater than some value exceeding 179°59′59″.
3. (a) Large since the radius, being 0.25 of a standard unit, is physically very great.
   (b) Yes, since the hyperbolic formula for the circumference of a circle becomes more nearly Euclidean as the radius decreases.
5. It is always sufficiently small.

CHAPTER VII

Section 4

5. Each side is $\pi r/2$ units long.
7. Consider a triangle with two vertices on the equator and one vertex at a pole.

Section 5

1. (a) Each pair of great circles meet in two points. Through each pair of points there passes at least one great circle. Through each point there pass infinitely many great circles, the totality of of whose points constitutes the sphere.
3. (a) It is the set of all points at a common distance from the same point.
   (b) It is a straight line; it is a point.
17. It decreases.

Section 7

Following Theorem 15:
1. No.
3. Let *A*, *B*, *C* divide their straight line into three equal parts.

CHAPTER VIII

Section 2

5. Definition the same as in VII, §5, Ex. 3(a). Yes.
7. No, since they meet in only one point.
9. (a) Solve $\cos c = \cos^2 \dfrac{\pi}{4}$ for c.

(b) Solve $\cos \dfrac{\pi}{10} = \cos^2 \dfrac{\pi}{10} + \sin^2 \dfrac{\pi}{10} \cos \lambda$ for λ.

11. (a) 2;   (b) 0;   (c) 1;   (d) 1.

Section 3

1. Definition the same as in VII, §5, Ex. 3(a).
3. The single elliptic plane.
5. (a) The set of all points whose distance from the center exceeds the radius.
   (b) Because there is no exterior.
9. No.
11. If a straight line is divided into 6 equal segments, then they are the desired unit segments.

Section 4

5. Any two straight lines are such circles, for they meet in just one point and a straight line is its own tangent.

Index